D0422281

What is Science for?

Bernard Dixon

What is Science for?

Harper & Row, Publishers
New York, Evanston, San Francisco, London

For information address Harper & Row, Publishers, Inc.,
10 East 53rd Street, New York, N.Y. 10022.

LIBRARY OF CONGRESS CATALOG CARD NUMBER: 73-1285

STANDARD BOOK NUMBER: 06-136 125-9

Contents

Preface

One of my pet dislikes is the book reviewer who demolishes the book an author *hasn't* written – the man who concedes that the parts on Beethoven, Mahler, and Schubert are passable but lambasts the writer for ignoring the important question of satellite communications. With this in mind, a plea: this is not intended to be (*a*) a book about the environment or contemporary disputation over global catastrophe (though there is an area of overlap here with my own field) or (*b*) yet another addition to the 'scholarly' literature on science and society. I am a journalist who happened to qualify in science some years ago and this book is my assessment of what science is and how scientists work; an account of the contemporary disquiet and criticism aimed at science and technology from both outside and inside the scientific community; and some suggestions about improvements in the way we decide how to support and utilise science. It is a book intended primarily for the general reader rather than the specialist.

In an attempt to reflect the contemporary scene, I have discussed the work of many other authors, and have given references to all such quotations and sources of information. This embellishes the book with academic trappings it does not deserve but will, I hope, be useful for those who wish to pursue particular topics further. With such carefully documented plagiarism, I have freed myself from every author's first conventional confession – that he stole all his ideas from his friends. The extent of my dependence on others – some friends, some not – is obvious. I must, however, enter the other usual disclaimer. None of those quoted are responsible for my use or misuse of their words and ideas. That is my responsibility alone.

Among other sources, I have also used some of my own material published previously in *New Scientist*, *The Spectator*, and *World Medicine*, and I am grateful to all three publications for permission to do so. I also gratefully acknowledge the following publishers for permission to reproduce material from the books named:

Penguin Books (*The Politics of American Science* by Dan Greenberg); Hamish Hamilton (*Scientists and War* by Lord Zuckerman); Ballantine Books Ltd (*The Population Bomb* by Paul R. Ehrlich); Macmillan (*As I Remember Him* by Hans Zinsser, *The Social Process of Innovation* by M. J. Mulkay, and *Wealth from Knowledge* by John Langrish *et al*); Heinemann (*The Art of Scientific Investigation* by W. I. B. Beveridge); the Clarendon Press, Oxford (*Scientific Knowledge and its Social Problems* by G. R. Ravetz); The Macmillan Company (*The Sociology of Science*, edited by B. Barber and W. Hirsch); Routledge and Kegan Paul (*Personal Knowledge* by Michael Polanyi); Cambridge University Press (*Public Knowledge: the Social Dimension of Science* by John Ziman and *The Foreseeable Future* by Sir George Thomson); Hutchinson (*Beyond Reductionism*, edited by Arthur Koestler and J. R. Smythies); Associated Scientific Publishers and the Ciba Foundation (*Civilisation and Science*); A. D. Peters and Company (*The New Anatomy of Britain* by Anthony Sampson, published by Hodder and Stoughton); Methuen (*The Art of the Soluble* by Sir Peter Medawar and *A Short History of Science* by J. G. Crowther); Yale University Press (*The Environmental Crisis* edited by H. E. Helfrich); Allen and Unwin (*The Body* by Anthony Smith); George Harrap (*Chemical and Biological Warfare*, edited by Steven Rose); Faber and Faber (*The Making of a Counter Culture* by Theodore Roszak); and Jonathan Cape (*The Technological Society* by Jacques Ellul, translated by John Wilkinson).

I am grateful to Sir Peter Medawar for permission to quote from his contribution to *Experiment* (edited by David Edge, BBC Publications); Professor Tom Cottrell for use of an extract from an article he first published in *The Listener*; Robin Clarke for his table contrasting 'soft technology' with the hard variety; Friends of the Earth for permission to reproduce an extract from their diary first published in *The Ecologist*; and HMSO for the use of extracts from the Rothschild report (Cmnd 4844). I also thank the publishers of the following publications for permission to use material as cited in references in the book: *British Medical Journal*, *Chemistry in Britain*, *Encounter*, *Impact*, *Journal of General Microbiology*, *Minerva*, *New Scientist*, *New Society*, *New York Times*, *Punch*,

Science, *Science Journal*, *Sunday Times*, *The Guardian*, *The Times*, *Times Literary Supplement*.

Finally, but most importantly, my thanks to Rena Feld, Gerald Leach, Tony Loftas, Martin Sherwood, Peter Stubbs, Nick Valéry, and Philip Ziegler for reading the original manuscript and intercepting many potential blunders; to Joe Hanlon and Michael Kenward for passing on valuable background information; to Joleen Huckle for scrupulous proof-reading; and to Katherine Adams, who typed every word of it.

Chapter 1
Ferment and uncertainty

Until very recently, anyone who asked the question 'What is science for?' could simply be categorised as foolish, provocative, or ignorant. Hadn't Francis Bacon explained triumphantly in the early seventeenth century that science was to be used for 'the merit and emolument of man'? Hadn't the civil engineers of the nineteenth century, and microbe hunters like Louis Pasteur, shown beyond doubt that science and technology were destined to be the chief architects of human progress? More recently, had not two world wars – the first fought with weapons devised by the chemists and the second won by the physicists – demonstrated the Promethean power of science over human affairs?

Yes, all these things were true. But now consider the most recent chapter in the story. Whatever may have happened to our political and social dreams of utopia, nineteenth century optimism as applied to science did not die when the century ended. It was, paradoxically, reinforced during the 1939–45 conflict, when scientists working on the Manhattan Project turned matter into energy and unleashed unprecedented power which helped to vanquish the forces of evil. If that were possible, science was capable of anything. In the years following the war, the developed countries were delighted to increase their research and development budgets each year, and to train more scientists, confident in the belief that science was a good thing, deserving of limitless support. In the United States, for example, government funding for science grew from 2·8 billion* dollars to 12·6 billion dollars per annum in the twelve years up to 1965. And the developing countries were only too anxious to follow.

Even as recently as 1961, Dr Alvin Weinberg, director of the Oak Ridge National Laboratory, Tennessee, summarised this

* billion is used throughout this book to mean 10^9 (one thousand million).

optimism by placing modern science firmly among Man's greatest artistic achievements: 'When history looks at the 20th century, she will see science and technology as its theme', he said. 'She will find in the monuments of Big Science – the huge rockets, the high-energy accelerators, the high-flux research reactors – symbols of our time just as surely as Notre Dame is a symbol of the Middle Ages.'[1]

Although those words do not specifically applaud scientific achievements, and could indeed be used to attack Man's devotion to science, there is no doubt about the confident mood in which they were written. Science was a great adventure of the human spirit, and a means of liberating man from thraldom, and we certainly needed more of it. The contrast with today's uncertainty and suspicion of science, and the crisis of confidence and identity now felt within the scientific community, is striking. It is illustrated most acutely in the way scientists themselves write about science. In 1945, Vannevar Bush was describing science in terms of *Endless Horizons*[2] and even as late as 1957, Sir George Thomson ended his book *The Foreseeable Future* with this flourish of boundless confidence: 'There is no reason to anticipate that anything irreparable will go wrong with the earth physically for many millions of years, and are there not other planets and other stars? . . . Even with the present brains of intelligent people Man may expect a glorious future. Who will dare to set limits to what he may achieve as his brain improves?'[3] In those days, few would dare to quibble with such sentiments. But the change since then has been drastic. In 1970, for example, another scientist, Dr Desmond King-Hele, began his *The End of the Twentieth Century*[4] by asking 'Will our civilisation destroy itself before the year 2000?' In place of buoyant, optimistic books on science, present-day attitudes to science are more accurately reflected by the titles of such volumes as *The Doomsday Book*,[5] *Can we Survive our Future?*,[6] *The Technological Threat*,[7] and *Science, Servant or Master?*[8]

In journals and magazines, too, the same trend is apparent. *Modern Science Wonder*, which I read avidly as a youngster, has long since disappeared. *New Scientist* magazine, launched in a mood of technomania in 1956 and gaining a valuable impetus from the first Russian sputniks, can no longer be as single-mindedly enthusiastic about science. Weinberg, writing in the American journal *Science*,[9] now finds it necessary to compose an article entitled 'In Defense of Science'. Harvey Brooks, Dean of the Division of Engineering and Applied Physics at Harvard University, and author of a recent influential report on science policy (see p. 211), asks 'Can science survive in the modern age?'.[10] William Bevan, the publisher of *Science*, writes anxiously about 'The Welfare of Science in an Era of Change'.[11] Mrs Shirley Williams, Minister of State in the Department of Education and Science in the British Labour government from 1967–69, writing in *The Times* on 27 February, 1971, announces that 'For the scientists, the party is over'.

Even as little as ten or twelve years ago, it was very different. For the intelligent young man, science was still one of those secure and obvious careers awaiting his talents, offering its rewards, and demanding his hard work and loyalty. As with accountancy, medicine, the law, and the church, science provided a satisfying social role and a defined career structure. Above all, by its very nature, it bestowed status in the community. Like the doctor, the cleric, and the teacher, the scientist was treated with natural respect as one with special knowledge and skills – as a person with a vitally necessary social role. Today, we still know that we require lawyers and accountants, just as we need the services of cobblers, airline pilots, dentists, builders, and taxi drivers. We may have doubts as to what they should teach and how they should teach it, but we go on training teachers because we continue to believe in education. We may blame doctors for iatrogenic disease, or criticise the ways in which some psychiatrists approach mental illness, but we know that medics are essential. The one practitioner we are not sure about is the scientist.

It all began with doubts in the middle 1960s, as report after report asked how many scientists we *really* needed and presented gloomy calculations about future reductions in the growth rate of investment in science. In part, this was an inevitable development. It was absurd to expect science to continue to grow at the dizzy pace of the post-war years and the 1950s: some deceleration was unavoidable. Lord Bowden, then Britain's Minister of State for Education and Science, expressed the dilemma in a speech delivered to the European Institute of Business Administration, held at Fontainebleau in September 1965: 'Processes which have gone on quite steadily for centuries may have to stop quite suddenly and quite soon, and . . . policies that we have been able to follow for decades and that we have never had to question before may have to be scrapped at great cost in spiritual anguish and perhaps at the price of great economic upheaval. Many of our traditional ideas have simultaneously and abruptly become obsolete, and we do not know how to adapt ourselves to the new situation.'[12] There were many who greeted such pronouncements with anger and disbelief, and strong words were bandied about urging that *any* reduction in expenditure on science could lead to unprecedented crisis. Since then, of course, the rate of growth in scientific expenditure has indeed fallen. Much dubious, low-grade, and unnecessary science has been lost in the process. But scientists both able and mediocre have found themselves out of work.

Another great change in recent years is that we are now much more concerned about the adverse consequences of our use of science and technology, from environmental pollution to hideous techniques of warfare. We are also more conscious of such potential social problems as malevolent exploitation of 'test-tube baby' experiments, invasion of privacy by computer, and clandestine mind-bending by psychosurgery and drugs. Numerous foolhardy developments, from the early 1960s onwards, have combined to reinforce public suspicions of the scientist as a sinister and irresponsible figure. On 9 July, 1962, for example, there was the American high altitude nuclear test – nicknamed the 'Rainbow bomb'. Against the warnings of Professor Fred Hoyle and other experts, and despite reassurances from US scientists, it produced an intense and long-lasting belt of radiation around the earth. A few months earlier, other research workers were putting millions of tiny copper wires into orbit round the earth, in the illusory hope of developing a communications system. There have been many similar examples since, from plans to move 27,000 tons of highly lethal nerve gas across the United States by train in the summer of 1969, to the type of dangerous and short-sighted misuse of pesticides highlighted by the campaigns of Rachel Carson. With air pollution increasing in many US cities, and lakes dying due to overwhelming pollution, concern burgeoned. Even *Time* magazine began an 'Environment' column in August 1969, as did many other newspapers and magazines. As the science historian J. G. Crowther pointed out early in 1972, whereas the drawings which Mayakovsky used to illustrate his poems had factory chimneys belching smoke as symbols of the triumphant virility of the Russian revolution, today smoke symbolises nothing but waste, destruction, greed, indifference, and disease.[13] And science, ultimately, is often the culprit.

Another major public theme of recent times is the Vietnam war, and here too science has played a massive role. US troops have used Indochina as a proving ground for the most sophisticated techniques of warfare ever conceived. Defoliation, employed on a huge scale in Vietnam to deny both food and cover to the enemy,[14] has already been copied in smaller conflicts in other parts of the world. South African mercenaries, for example, waged chemical warfare to help the Portuguese fight nationalist guerrillas during 1972 in the jungles of northern Mozambique.[15] The ecological effects of such warfare – and indeed of the 26 million bomb and shell craters estimated to have been caused by 'conventional warfare' in Indochina[16] – will be extremely long-lasting, but the full extent of the damage will not be known for many years.

Such happenings have introduced a new fear of positive evil into public contemplation about science. There is a contrast here with Arthur Koestler's description of the first generation of modern scientists – Kepler, Galileo, and Newton – as *The Sleepwalkers*, men who were exploring entirely new fields without realising what the future held, and with the view of Lord Rutherford typified by a description of him in the announcement for a television programme in October 1971: 'A simple, practical man, he made his apparatus out of petrol pumps and tea chests. He hummed "Onward Christian Soldiers" as he performed and inspired work which led to the atomic bomb.'[17] The uncertain dangers broached by the pioneer can be forgiven, particularly if he is a historic figure. Not so the scientist who directs his skills to 'improving' napalm so that it burns the skin of Vietnamese peasants more efficiently. James Cameron, writing in an issue of *Punch*[18] devoted to science, has most accurately captured the non-scientist's dread of science in warfare. He is discussing his visit to Bikini Atoll in the summer of 1946 to watch an atomic test: 'I began to mistrust scientists at just this point, when they explained that they had knocked off 100,000-odd people in one twenty-sixth part of a second without exactly knowing how. That was why we were trailing across from California to the Marshall Islands to do it all again, in laboratory conditions.' It is such feelings, nurtured in recent years by scientists' involvement in the Vietnam conflict, which have led to impatience and anger with those scientists who try to pretend that their work is somehow separate from the rest of their lives. As a result, scientists have found themselves facing an unexpectedly hostile public. When Professor Murray Gell-Mann, the Nobel prize-winning physicist, stood up to speak at a meeting in the Collège de France in Paris in June 1972, for example, his audience included a large number of people who wanted him to talk not about his special subject of subatomic particles, but about Vietnam. Gell-Mann had previously worked in defence,[19] and his critics were understandably furious that he was unwilling to comment on scientific warfare in Vietnam.

As well as the negative and destructive effects of science, another reason for the changed climate of thinking is disappointment at the hard returns from science which has, at times in the past, promised more than it has delivered. We have realised that science is not the supremely important tool we once thought it to be for improving the world, and that other activities may be more productive and more deserving areas for financial support. We now know, for example, that there is little convincing evidence of a positive association between investment in science and the economic growth of companies and countries (pp 104, 115); less exalted skills, such as those of the marketing man and the manager, are at least as significant as those of the scientist. Even medicine, which in the past has been transformed out of all recognition by the advent of science, has had to face the possibility of a new ordering of priorities. There are fears that the best doctors have become so preoccupied with the frontiers of sophisticated science-based medicine that they neglect more mundane but more widespread problems. The United States, for example, though pioneering many dazzling frontiers of medical research, has fallen from being world leader to fifteenth place in infant mortality rates – one of the most reliable indexes of medical standards. Most recently, on a variety of fronts from the philosophical to the ruggedly practical, science has been under attack for the sort of world it has produced. The whole basis of our science-based, growth-obsessed industrial civilisation is being questioned. As one of the chief architects of that society, the scientist is now on the spot.

To their credit, a few scientists have played an active part in publicising these issues and in promoting the debate about the future directions taken by society and the role of science within it. (We shall cover some of their main points and activities later in this book.) But many scientists have not. Instead, they think, talk, and feel as before, unable to comprehend the radically changed circumstances in which they pursue their craft. Why have scientists taken the change so badly? Clergymen, for example, have been through a comparable state of doubt and ferment in recent years, and have emerged less hurt and unhappy than many of today's scientists. The crisis for clerics began some years before that facing scientists, but otherwise the two situations are very similar. One clue to the scientists' greater unease may be found in the uniqueness of their former status. People in most trades and professions can be categorised in one of two ways: the pillars of society, whose work supports and sustains the community; and the innovators and rebels who propel it forward. Scientists like to think of themselves as both. They serve as technicians, oiling the wheels of society, commerce, and industry, and keeping the great machine working, with its increasingly intricate internal interconnections. They also appear as magicians, producing a copious flow of new products and processes used for everything from generating power or feeding the hungry to assessing our legal right to drive after drinking or determining the sex of Olympic athletes.

Suddenly, scientists are under scrutiny on both fronts. While the full-time creators – the poets and the artists and the composers – carry on doing their thing in a secure (even if poorly paid) social niche, and while the plumbers continue to plumb pipes and the doctors cure disease, scientists face growing doubts about both of their roles in society. We are not sure we like our technocratic civilisation, and we doubt whether we want more and better gadgets and cleverer theories to take us further along the same road. As I shall argue (p. 219), whatever happens in future we shall continue to need science and scientists, if only to resolve the problems they have created. But at the moment scientists are passing through a corporate menopause, stricken by anxiety as to how they will emerge from it. Some will surface unscathed and even rejuvenated from the process. Others won't.

Sociologists' neglect of science

It is odd that, until recently, sociologists have paid little attention to science and scientists as social phenomena. During the inexorable growth of science from the seventeenth century onwards, there was virtually no attempt to describe and analyse the workings of the scientific community. Only since the early 1960s – after the emergence of science as an organised profession, and indeed during a period of recession in the financial support for science – have sociologists begun to take notice. This is not merely a reflection of the growth of sociology itself; while happily poking their noses into every other aspect of society, they have been genuinely reluctant to scrutinise science. Some of the reasons are clear enough – the intimidating complexity and jargon of science, the suffocatingly intellectual and unreal flavour of the academic discipline known as the 'philosophy of science', and the pathological exclusion of the real stuff of scientific intercourse from scientific journals and papers. But the sociologists' neglect of science is still a surprising omission, and one we could live to regret. Now that the party really is over for scientists, there is a growing belief (separate from, but drawing additional fire from, the environmental damage wrought by science and technology) that science has over-sold itself, and understandable resentment that the scientific community has been allowed to grow up in our midst into an autonomous priesthood upon which the citizen has little or no influence. More open discussion of science, its internal conduct and external consequences, at an earlier stage would have helped to forestall this development.

As it is, science today confronts a situation of acute crisis and uncertainty. On the one hand it still consumes a considerable amount of money – the US government supported research and development to the tune of 16·7 billion dollars in 1972 – and at the other extreme we hear influential critics calling for moratoria on research and in some cases asking whether science is a worthwhile pursuit at all. In confronting the question 'What is science for?' in this book, I hope to reflect this contemporary ferment, as well as presenting some ideas about desirable changes in the way we support and exploit science. But first we must consider the prior question 'What is science?' Contrary to expectations, it is one which is not open to a neat and tidy answer.

Chapter 2
What is science?

Ask half a dozen accountants, butchers, or lawyers to describe how they go about their daily work, and they will have little difficulty. The members of each group will scarcely disagree about the way in which their crafts are practised. Even people outside these occupations will be able to explain reasonably accurately the skills used by the various practitioners.

Not so science. To the outsider, acquainted with fictional scientists in literature, television, and films, and most conscious of real scientists as purveyors of bombs and pollution or spurious authorities used in advertising soap powder, scientists appear in a variety of guises – remote, eccentric, magical, dangerous, naive. But there is unquestioned acceptance that scientists tackle their research work by applying an unfamiliar, possibly mysterious, and certainly superior brand of mental process. This is known as *the* scientific method. On the other hand, scientists themselves are far from agreed about the intellectual basis of their work. Most of them never bother to think about it, leaving such speculation to those earnest gentlemen, the philosophers of science, who write inscrutable books on the subject.

Considering the lengthy history of science, and its enormous capacity to change the world, this is a bizarre state of affairs. If we are to take a more intelligent, active interest in the application of science in society, we ought to understand how scientists pursue their craft. What, then, is a scientist? Is there, indeed, such a definable creature? In trying to answer these questions, it is sensible to ignore for the moment what the professional philosophers of science say. The most reliable approach is simply to observe the everyday work of people who call themselves scientists. What we find when we do this is an unexpected profusion of activities, so different as to throw into doubt the usefulness of terms as wide as 'science' and 'scientists'. At its extremes, such scrutiny exposes both the truly mysterious creativity of the outstanding scientific thinker, and the humdrum banality of research that differs little from the action of a child collecting car numbers.

Consider five examples, typical of the sort of science now being pursued in the advanced Western countries. First, there is Dr A, who works as a microbiologist in a large pharmaceutical company. He heads a team of thirty people testing chemical substances in the hope of coming up with new drugs to combat virus infections, such as influenza and the common cold – drugs comparable with the antibiotics used against bacterial diseases. Another team of scientists, working alongside, consists mainly of chemists who synthesise likely substances and hand them over for screening. Dr A's job is to supervise the technicians and graduate scientists under his charge as they systematically test the activity of an endless stream of new substances against different viruses. For the most part, this involves adding measured quantities of the substances to specific viruses multiplying in cells growing in glass vessels, and measuring the diminution of virus growth, if any. Of prime importance to Dr A is the reliability and standardisation of these methods. He must be sure that the tests are as sensitive as possible, in order to reveal even the tiniest effects – which could be significant in suggesting ways of altering a particular substance for greater activity. Moreover, there must be no variations in the materials or manipulations employed in the tests from one day to the next, or between one technician and another. Dr A also requires a thorough knowledge of statistics, which he uses to process the results. The entire project hinges on scrupulous standardisation of test conditions, and the ability to detect a vanishingly small positive effect in any one of the hundreds of thousands of different tests conducted each week.

Now consider Dr B, a theoretical physicist on the staff of a university department. He works on what is ostensibly one of the most materialistic of scientific disciplines, subatomic physics, in an attempt to learn more about the structure of matter. Yet he spends most of his time sitting at a desk, thinking and calculating, and consulting books and journals in the library. Unlike Dr A, he never touches a test-tube, oscilloscope, or other piece of experimental equipment; but he does use a giant computer in the university, booking his share of time on it each week. Dr B is interested in the structure of the nucleus of the atom. At one time the nucleus was thought to be a fundamental unit of matter, but we now know that it has a complex structure of its own. Beginning with measurements that have been made with particle accelerators in the more practical world of high energy physics, Dr B tries to work out how certain forces inside the nucleus account for its physical behaviour. There is, he feels, a serious theoretical gap in our understanding here. Just as worrying is the irritating complexity of present views on subatomic structure. Because nature is usually both elegant and simple, these ideas are probably mistaken. But the answer to Dr B's puzzle is unlikely to come simply by doing one of his weekly sums on the computer. It is just as likely that the computer will be used only to *confirm* an imaginative idea dreamed-up one day by Dr B or one of his colleagues. As with Archimedes and his cry of 'Eureka!', that is the way in which many major break-throughs in physics have come in the past.

More like a young schoolboy's idea of science is the work of Dr C, a green-fingered scientist who spends much of his professional time (and much of his home life too) inventing gadgets. He works for a firm of consulting scientists who solve problems brought to them by small manufacturing companies who cannot afford to support scientific staffs of their own. Most of the problems are thoroughly practical – such as how to prevent moulds growing on a new type of plasterboard, or how to detect errors in the operation of a semi-automatic production line making mechanical toys. Dr C's workbench reflects this mixed bag of interests, and is littered with wire, pieces of strip metal, meters of various sorts, glassware, and other odds and ends. On a typical day, he is to be found fiddling with a light-sensitive photo-cell, which he is mounting inside a heat-resistant shield for operation inside a furnace used at a nearby iron-works. It is part of a new system of temperature monitoring, and is the third adaptation of a design perfected by Dr C some months ago. He has had the gadget patented, and it will shortly be manufactured by an associated company on a commercial basis.

More difficult to categorise is Dr D, a biochemist who recently won a Nobel prize for his work. His aim is to discover how living cells regulate their metabolism. In the first half of this century, biologists studied the structure of cells and began to compile detailed accounts of the various chemical substances they were made of, and how they lived by releasing energy from foodstuffs and building up new cell material. Since then, there has been increasing interest in how this complicated machinery is regulated. How, for example, do the various parts work together so harmoniously – a harmony disastrously destroyed in the cancer cell? Dr D's contribution has been to piece together, like a jigsaw, evidence drawn from many different sorts of scientific inquiry – from chemical analysis of components of the cell to genetic studies on how bacteria inherit the capacity to resist destruction by antibiotics. He himself is part theoretician and thinker, part experimenter. The end result has been a series of discoveries on the cell's control mechanisms, made by testing laboriously speculative ideas on how the cell functions. Dr D is in many ways a hybrid of Drs A, B, and C, applying logic, systematic work, and practical inventiveness to the testing of bright ideas.

For Dr E, my final example, science is a matter of following cookery recipes. Like all scientists, he learned a lot of scientific facts during his training as a physiologist, and he continues to read widely, keeping up to date with the latest research publications. But he is virtually impotent as an innovator, devoid of imaginative creativity, logical gymnastics, and practical dexterity. So he simply copies what other people do. He will, for example, notice that another physiologist has used such-and-such a method to measure how the kidney works in the hamster, and decide to try the same technique in the rabbit. He acquires some rabbits, obtains a grant for purchasing the necessary equipment, and does exactly what the other scientist did. The project yields a series of measurements which he writes up for publication, following closely the structure of the other man's paper. It is all competently done, but Dr E has brought nothing new to the problem. His results are reliable and, because they have never been obtained before, interesting. Someone else, some day, may use them in making real progress. But for the moment Dr E has to move on to another project, because he doesn't know what to do next.

There are some common elements in these five stories. The inventor, for example, has occasional imaginative ideas that help him conquer knotty practical problems, while Dr A sometimes devises new pieces of equipment to increase the sensitivity of a test or the rate at which samples of potential new drugs are tested. But the fundamental character of the work in each case is different. These scientists succeed for very different reasons. Moreover, while I have chosen to use five competent scientists as my subjects, many research workers have lesser talents. There are, for example, bread-and-butter investigators like Dr E who are not even conscientious, and so clog up scientific literature with unreliable data.

What the philosophers say

At this point, let us turn to what the philosophers, commentators, and historians of science have to say about scientists and their work (concentrating on those who have actually *done* science rather than just written about it). Broadly speaking, there are two conceptions of science. As eloquently described by Sir Peter Medawar in *The Art of the Soluble*[1] these are the romantic and poetical, or the rational and analytical, views. The first is based primarily on the idea of imaginative insight, the second on the evidence of the senses. Of the first, Medawar writes, 'truth takes shape in the mind of the observer: it is his imaginative grasp of *what might be true* that provides the incentive for finding out, so far as he can, what *is* true. Every advance in science is therefore the outcome of a speculative adventure, an excursion into the unknown. According to the opposite view, truth resides in nature and is to be got at only through the evidence of the senses: apprehension leads by a direct pathway to comprehension, and the scientist's task is essentially one of *discernment*.'

Such rigid distinctions are not satisfactory. In practice, the two interpretations tend rather to go with different temperaments - much as Tory and Labour, or Republican and Democrat, are used as labels for certain complexes of opinion. They also generate a similar amount of anger when their protagonists come into conflict, each convinced that his own view is correct.

Both, in fact, are partly right. There are numerous episodes in the history of science which compel one to believe that, at times, scientific discovery is comparable with artistic creativity. Time and time again, scientists have written about their work in terms of intuition or sudden, unexpected insight. Professor Dennis Gabor, who won the Nobel Prize for Physics in 1971, said that the idea for his crucial discovery of holography occurred to him not while working in the laboratory or library but while sitting on a bench watching tennis. It was the most exciting moment of his life. Most of his ideas, he said, came to him while shaving. Such was the way with the chemist Friedrich Kekulé who, in 1865, dreamed up the way in which carbon atoms are linked together in the benzene molecule. Writing about it afterwards, he tells how he was dozing in front of the fire when suddenly he saw carbon atoms dancing:

'My mental vision rendered more acute by repeated visions of the kind, could now distinguish larger structures of manifold confirmation: long rows, sometimes more closely fitted together, all twining and twisting in snake-like motion. But look! What was that? One of the snakes had seized hold of its own tail, and the form whirled mockingly before my eyes. As if by a flash of lightning I awoke: and this time also I spent the rest of the night working out the consequences of the hypothesis.'[2]

It is true, of course, that Kekulé had been mulling over his problem for some months, and he had to apply his intellect to the task of verifying his imaginative solution and studying its logical consequences. But the crucial idea emerged in a fashion more romantic than rational. 'There is no logical way to the discovery of these elemental laws,' wrote Albert Einstein about the physical laws of the universe.[3] 'There is only the way of intuition, which is helped by a feeling for the order lying behind the appearance.'

Arthur Koestler is the best-known of contemporary protagonists of the mysterious, emotional view of scientific creativity. But he has also tried to analyse the phenomenon in psychological terms. In his book *The Act of Creation*,[4] Koestler sets out his theory of the 'bisociative act' – the vision whereby we see a familiar object from a new aspect. This is common, he suggests, to the appreciation of jokes, to scientific discovery and to artistic creation. We laugh when an event we previously believed to belong in one context is suddenly seen to belong to another – when we see that the fat man we previously saw as a threat or rival has slipped on a banana skin. The scientist, too, at the point of discovery, perceives a previously hidden analogy.

When we move towards the rational concept of scientific

discovery, a morass of learned discussion awaits us on the library shelves. Theory after theory has been proposed and dissected in dusty tomes. One major slab, which is now generally condemned as ill-founded, advocates the process of *induction* as the basis of science. The idea, which began with Francis Bacon, is that scientists operate by collecting large numbers of facts, from which conclusions somehow emerge. Simply by confronting lots of sense observations, garnered without preconceived ideas or prejudice, the scientist can discern new relationships, new laws of nature. The mathematician Karl Pearson was the greatest protagonist of induction, and probably one of the last working scientists to believe in it as a universal conception of science. In his book *The Grammar of Science*,[5] he claims that 'the classification of facts, and the formulation of absolute judgements upon the basis of this classification – judgements independent of the idiosyncrasies of the modern mind – is peculiarly the scope and method of modern science.'

Today, it is routine among professional scientists to say that *of course* this is not how science proceeds. Sir Peter Medawar, who has such disarming insight into these things, even disputes what Charles Darwin said about his own work in a famous passage in his autobiography. 'I worked on true Baconian principles,' wrote Darwin, 'and without any theory collected facts on a wholesale scale, by printed enquiries, by conversation with skilful breeders and gardeners, and by extensive reading.' Not at all, said Sir Peter in a radio talk in 1967, 'he did not reason in this way at all.'[6]

The conventional view of scientific discovery nowadays is the 'hypothetico-deductive' system, immortalised in Sir Karl Popper's *Logik der Forschung*,[7] published in 1934. In part, this reconciles the romantic and rational schools of thought. The scientist is seen as a creature who alternates, sometimes rapidly, between imaginative and critical phases of thought. During the imaginative period, he makes a guess about some aspect of the world, and frames a hypothesis. Then he subjects his imaginative speculation to ruthless criticism. By deduction and experiment he tries to falsify (disprove) his own hypothesis. Only when the hypothesis has survived severe scrutiny can it be even temporarily accepted. The two phases must be kept in check. 'During the idea-producing phase of the process,' writes the statistician Leo Törnqvist, 'it might be dangerous to let critical thinking play such an important role that it blocks the flow of promising ideas, but some control is needed for avoiding the possibility of being caught too early by one idea when another maybe clearly better one still waits to be detected.'[8]

There is little doubt that the hypothetico-deductive interpretation of science is a valid one. But it is also something of an élitist view of some of the *best science*. Indeed, its historical exponents (including the American philosopher C. S. Peirce, the British economist Stanley Jevons, the French physiologist Claude Bernard, and the British scientist William Whewell) and its leading protagonists today (of which Medawar is the best known) are all men of high intellectual calibre. Alas, much of what goes on inside research laboratories today simply does not merit the label of 'hypothetico-deductive' – or any other grand-sounding title.

This was brought out in a broadcast talk[9] in 1960, in which Professor Tom Cottrell launched a strong criticism of much professional science, on two main counts:

'The first is that the average chemist, in particular, is reluctant to master and apply scientific theory. So often, when faced with a problem he has not met before, he resorts to the purest empiricism. For him an education in theory is equivalent to the education in Latin that the scientist attending some universities must have undergone: it is something to be endured for the sake of getting a degree. The second defect is an insufficiently critical and inquiring attitude to the research problem itself. Much effort is spent investigating problems that do not require solution, and which could have been shown by a careful preliminary analysis not to require solution.'

Another fact which distinguished commentators tend to overlook is that the various approaches to science – and I hope I have shown that no single descriptive epithet is adequate – are to some extent alternative ways of attaining the same goal. Introducing his *Scientists and War*,[10] Sir Solly (now Lord) Zuckerman argues that, just as King Canute learned that he could not turn back the sea, so 'the layman' cannot command a scientist or technologist to make a breakthrough in a particular field: 'A cure for cancer will be discovered only when some new genius reaches a new understanding; one day perhaps another genius will discover how to tame the processes of nuclear fusion, in the same way as that of fission has been; and in the fullness of time maybe someone will develop a simple and effective device which will allow us to see in the dark. But none of these things will happen except as the spontaneously creative acts of particular gifted individuals.'

Yet – as the story of discoveries made under the pressure of wartime necessity show – the processes of innovation and research *can* be hastened. Discoveries which might emerge as creative acts of gifted individuals can be forced out of recalcitrant matter simply by bringing more scientists, more techniques, more money, to bear. This may be an inelegant and uneconomical way of tackling problems, but it works. Like those statistical monkeys hammering out the sonnets of Shakespeare on their typewriters, large scale deployment of competence can compensate for the lack of the single man of genius. The American sociologist Professor Robert Merton, who has made a special study of multiple discovery of the same phenomenon by different scientists, even argues that scientific geniuses typically make discoveries that are later made independently by several scientists of lesser talent.[11]

Two other factors have a profound effect on the conduct of scientific research – practical innovation in techniques and equipment, and pure luck. Again, the philosophers of science neglect both. Firstly, they appear to be out of touch with the actual stuff of research – with test-tubes and microscopes and the sweaty drudgery of much routine benchwork. They tend to be totally intellectual and to overlook the fact that scientific progress is catalysed almost as often by methods of doing things as by inspired or logical thought. Practical techniques and novel pieces of equipment are to be found littered around the genealogical tree of science, especially around its major turning points. A new micromanipulator, for example, which allows biologists to probe cells with greater delicacy than ever before, or a way of using radioactive isotopes to chart previously inaccessible regions of living cells, are likely to open up entirely new realms of science. Two developments in recent years which automatically spawned massive new lines of investigation were the electron microscope and the radio telescope.

When medical research workers look back at the end of the

century, they will probably select as one of the most important single developments of the 1960s techniques for growing (and hence studying) human cells in the laboratory. As we shall see in Chapter 11, this work is fraught with major implications, for both good and evil. Biologists can now grow in laboratory glassware cells containing an individual person's genetic material, the DNA, which directs the development of the entire body. A microbial version of John Smith can be cultivated, like a bacterium, in the laboratory. John Smith's hereditary material can thus be studied with the same facility that would be possible if John Smith produced offspring numbered in thousands, which would allow conventional genetic studies to be conducted.

Several other possibilities, some of them bizarre, pose potentially serious problems. Others are undoubtedly for the good. One of the most important applications of human cell culture is in studying certain inherited diseases. Already, several of these conditions, including maple syrup urine disease and cystic fibrosis, have been found to manifest themselves in cultured cells. Such cells can thus be used to diagnose the conditions at an early stage (when treatment may be more effective) and to investigate the disease process itself. Finally, there is the possibility of curing such inherited conditions by 'genetic engineering', by tinkering with the hereditary material inside a patient's cells rather than simply making the best of the bodily deficiencies, as at present. When the new science of genetic engineering becomes a routine procedure its success will owe much to the teams of research workers who have devised just the right conditions for growing human cells in the laboratory. Because of its importance, such work has now assumed the form of 'mission-orientated' research, particularly in the United States where it is being tackled by large teams of scientists.

At the opposite pole to the large scale research programme is one of the classical stories that illustrates the seminal importance of chance observation in science – Alexander Fleming's discovery of penicillin. Recent investigations of this historic incident, conducted by Dr Ronald Hare and described in his book *The Birth of Penicillin and the Disarming of Microbes*,[12] show that the episode was even more remarkable than we had been led to believe by Fleming himself.

Briefly, the accepted version of the penicillin saga is as follows.

One summer day in 1928 a spore of the mould *Penicillium* drifted through a window at St Mary's Hospital in London, where a meticulous Scots bacteriologist was studying cultures of staphylococci – the type of bacteria that cause boils. The mould happened to land on one of his glass dishes, exposed briefly to the air during examination. Days later, with the experiment completed, Fleming picked up a discarded culture to show to a colleague. He noticed that it had been contaminated by the mould, which was growing alongside the tiny colonies of staphylococci, and causing them to become translucent. As Fleming later recorded, 'what had originally been a well-grown staphylococcal colony was now a faint shadow of its former self'. Though such 'antagonism' between microbes is common on old, contaminated cultures, Fleming grasped its significance for the first time – the mould was producing a substance that had diffused into the medium and was attacking the bacteria. Though it was to be another twelve years before penicillin was introduced into medical treatment, Fleming realised the great potential importance of his substance in treating infectious disease. The discovery was a stroke of luck but, as Louis Pasteur, the pioneer French microbiologist said almost a century earlier, 'chance favours the prepared mind'.

What Ronald Hare has shown is that the odds against the episode were much greater than was thought previously, and that the historic contaminated plate could not have originated by the accepted sequence of events. When he retired in 1964, Dr Hare set out to 'rediscover' penicillin by recreating the original conditions. Despite using considerable ingenuity, this proved impossible; it was extremely difficult even to cause the mould to grow on a culture of staphylococci. The reason is simple. As we now know, penicillin acts only on growing bacteria. It cannot, as Fleming thought, attack a mature colony.

The only possible explanation is that Fleming's mould landed on the plate while he was inoculating it with staphylococci and that he never incubated the plate in the usual way. If the plate had then been left on the bench, at a relatively low temperature, the mould could have had an opportunity to grow appreciably and produce penicillin *before* the staphylococci started multiplying – at which point growing colonies in contact with the antibiotic would be affected by it. Following an intriguing piece of detective work, aided by the British Meteorological Office's records of summer temperatures for 1928, Hare proposes that this is what actually happened.

Moreover, it seems that Fleming's window was rarely if ever opened. It is far more likely that the mould spore came from the laboratory immediately below Fleming's room, where C. J. La Touche was studying moulds. When Fleming later tested La Touche's moulds, of his eight strains of *Penicillium* only one produced a measurable amount of penicillin. It turned out to be a most unusual strain, whose high yield of penicillin could be matched by only two others harvested from a world-wide search conducted by American research workers some years later.

Scientists are usually reluctant to admit the importance of luck in their work. They prefer to pretend that their discoveries follow lofty flights of reasoning – and are encouraged to do so by the very methods by which they communicate their results to other scientists, as we shall see in the next chapter. Occasionally, however, a scientist does tell all. One intriguing example[13] is the way in which the Russian endocrinologist Andrew Nalbandov, working in the United States, discovered a method of keeping experimental chickens alive after hypophysectomy – an operation to remove the pituitary gland:

'In 1940 I became interested in the effects of hypophysectomy on chickens. After I had mastered the surgical technique, my birds continued to die and within a few weeks after the operation none remained alive. Neither replacement therapy nor any other precautions taken helped and I was about ready to agree with A. S. Parkes and R. T. Hill, who had done similar operations in England, that hypophysectomised chickens simply cannot live. I resigned myself to doing a few shorter experiments and dropping the whole project when suddenly 98% of a group of hypophysectomised birds survived for three weeks and a great many lived for as long as six months. The only explanation I could find was that my surgical technique had improved with practice. At about this time, and when I was ready to start a long-term experiment, the birds again started dying and within a week both recently-operated birds, and those which had lived for several months, were dead. This, of course, argued against surgical proficiency. I continued with the project since I now knew that they could live under some circumstances which, however, eluded me completely. At about this time I had a second successful period during which the mortality was very low. But, despite careful analysis of records (the possibility of disease and many other factors were considered and eliminated), no explanation was apparent. You can imagine how frustrating it was to be unable to take advantage of something that was obviously having a profound effect on the ability of these animals to withstand the operation.

'Late one night I was driving home from a party via a road which passes the laboratory. Even though it was 2.00 a.m., lights were burning in the animal rooms. I thought that a careless student had left them on so I stopped to turn them off. A few nights later I noted again that lights had been left on all night. Upon inquiry it turned out that a substitute janitor, whose job it was to make sure at midnight that all the windows were closed and doors locked, preferred to leave on the light in the animal house in order to be able to find the exit door (the light switches not being near the door). Further checking showed that the two survival periods coincided with the time when the substitute janitor was on the job. Controlled experiments soon showed that hypophysectomised chickens kept in darkness all died, while chickens lighted for two one-hour periods nightly lived indefinitely. The explanation was that birds in the dark do not eat and develop hypoglycaemia [a low level of sugar in the bloodstream] from which they cannot recover, while birds which are lighted eat enough to prevent hypoglycaemia. Since that time we no longer experience any trouble in maintaining hypophysectomised birds for as long as we wish.'

Ignorance, like luck, is another factor that commentators frequently neglect as a contribution to scientific discovery. Indeed, the two are often linked together, in the case of the scientist who stumbles on a new phenomenon by doing experiments which conventional wisdom would have ruled out of court. The pioneer biochemist Sir Frederick Gowland Hopkins once remarked that, 'The advantage of a certain amount of ignorance is that it keeps you from knowing why what you have just observed could not have happened.' Henry Ford put it more succinctly: 'An expert is someone who tells you that things cannot be done.' Sometimes a new discovery is made when an apparent barrier, part of the orthodox theoretical structure of science, is flouted by a scientist imaginative and bold enough to question the obvious. More often, in such cases, there is a mixture of chance and ignorance.

The picture which emerges from our study of how scientists work is of a variety of different methods and attitudes. The problem simply isn't resolvable by giving science a single descriptive epithet – 'hypothetico-deductive', or 'logico-inductive' (another favourite), or 'logical inferences from empirical observations' (yet another). There *are* scientists who stumble along, as Tom Cottrell says, by blind empiricism, which is wasteful in time and money. There *are* huge research programmes based on wholesale induction. (The screening project for anti-tumour drugs, run in recent years at the US National Institutes of Health at Bethesda, is one example, in which hundreds of thousands of different preparations have been tested.) And there *are* scientists who have flashes of intuitive insight which they themselves do not understand. Despite the fact that science, once a pursuit for the leisured amateur, is now a highly organised profession, it embraces a variety of very different approaches.

Is there, in that case, such a thing as *the* scientific method? Certainly, many books have been written arguing that there is and that it should be applied more widely to social problems, outside conventional research. One of the most important this century was Professor C. H. Waddington's *The Scientific Attitude*,[14] first published in 1941, in which he vigorously urged the value of scientific thinking as against 'the emptiness of fascism'. Science, he said, was not just a collection of tricks. It was 'an attitude to the world, a way of living'.

What Waddington had in mind was a certain rigour of thought and reflection about problems, in contrast to ill-informed guesswork or prejudice. Whether testing an imaginative idea or running a screening programme for new drugs, the scientist applies exactitude in defining his problem and collecting and interpreting his data. It is this that is common to the various brands of science I have described. The tactics are the same even when the strategy varies greatly. But whether it should be called 'scientific method' is another matter, because on careful scrutiny scientific method turns out to be little or no different from the analytical processes used in many other jobs and professions – by a detective seeking a murderer, for example. It amounts to no more than scrupulously applied common sense.

A classical example of this critical temper is to be found in Sir Francis Galton's 'Statistical Inquiries into the Efficacy of Prayer', published in 1872.[15] Trying to discover whether or not prayer works, Galton simply set out to determine whether members of the British royal family, who are 'prayed for' every day in churches throughout Britain, live any longer than commoners as a result of these national supplications for their wellbeing. His statistical analysis showed that, if anything, they fared worse than professional people of humble birth. However, Galton reasoned, the gross amount of prayer for the royal family cannot be assumed to be proportional to its sincerity. No-one, on the other hand, could doubt the sincerity of prayers for the lives of newborn children. So he decided to find out whether stillbirths were any less common among the children of the devout than among the children of professional people generally. Statistical analysis showed the frequency to be about the same in the two groups.

In this work, Galton was not just poking fun at religion. He acknowledged that prayer could strengthen resolution and bring serenity in distress, and claimed that he did not 'profess to throw light on the question of how far it is possible for man to commune in his heart with God'. What Galton was doing was to scrutinise rigorously and factually the question of whether prayer achieves a defined object. This he did by 'scientific method', and his study throws into prominence its recurrent feature – that of strictly defining the problem and the conclusions drawn. It is at this level of methodology that all scientists find common ground. Of central importance to the tactics of science are such procedures as statistical analysis of results in a search for significance, and the 'controlled' experiment, meaning one in which usually only one factor is allowed to change at a time, all other conditions being kept constant, in order that the experimenter can assess the effect of the changing factor.[16]

Even in the tactical application of these scientific tools, however, the enormous contrast between the outstanding creative scientist and his bread-and-butter colleagues is exposed. Where the blind empiricist will stumble from one question to the next, and the scientist with the accountant mentality will deploy a massive induction-based experiment (which may somehow throw up the result he is looking for), the top-class creative thinker designs a single, crucial experiment that decides absolutely between one hypothesis and another. Hans Zinsser recorded this talent in the great French bacteriologist Charles Nicolle. 'Nicolle did relatively few and simple experiments,' he said. 'But every time he did one, it was the result of long hours of intellectual incubation, during which all possible variants had been considered and were allowed for in the final tests. Then he went straight to the point, without wasted motion. That was the method of Pasteur, as it has been of all the really great men of our calling, whose simple, conclusive experiments are a joy to those able to appreciate them.'[17] On the other hand, as Ortega y Gasset observed, 'contemporary science, with its systems and methods, can put blockheads to good use'.[18]

We began this chapter with some of the clichés commonly used to describe 'the scientist'. One attribute common to the various caricatures of the working scientist, whether portrayed as a dotty eccentric or sinister megalomaniac, is solitariness. We tend to see scientists as remote, driven into isolation by the pioneering nature of their occupation, and the high intellectual demands it makes upon them. Of all the misunderstandings about scientists and how they work, this is the most pernicious. It is dangerous because it obscures the most valuable aspect of scientific inquiry – its continual momentum towards self-correction, which propels scientific interpretations of the world more and more closely towards the trustworthy and reliable – towards what we call 'the truth'. This happens because science is not a pursuit for the lone recluse, but an activity of an international community of research workers who constantly cross-check their findings and criticise each others' work. That is what we shall consider in the next chapter.

Chapter 3
The scientific community

The lone scientist, withdrawn from the world, is a haunting image. Newton sitting alone in a country orchard, Einstein brooding over curved space while standing on a bridge looking into the water – these are well-known cameos of great scientists at their most creative moments. They are also totally misleading as pictures of how science really works. The scientific genius may have his flashes of inspiration while in solitary, but if he is to be an effective scientist he must be very much part of the community of science, not withdrawn from it. 'The scientist, however remote he may seem, is always bound closely to the scientific life around him,' writes Sir Alan Cottrell.[1] 'He cannot work in a vacuum. He has to take the ideas and problems as they exist among his fellows, transmute them in his own personal way, and then bring them back as offerings to his community. He both takes and gives, in the scientific currency of his time.'

This is so because science is an evolutionary subject, a massive and intricately cross-connected activity that owes its success to the honesty and openness of its practitioners in comparing and collating their work. Many human activities have their meetings and publications and international associations, but such social intercourse is a vital and inseparable part of science. As Professor John Ziman argues in his book *Public Knowledge, The Social Dimension of Science*,[2] this dimension is neglected in the conventional philosophical approach to science. 'Every scientist sees through his own eyes – and also through the eyes of his predecessors and colleagues. It is never one individual that goes through all the steps of the logico-inductive chain; it is a group of individuals, dividing their labour but continuously and jealously checking each other's contributions.'

A typical example of a long, unpredictable, and extremely mixed chain of discovery is that leading to our current knowledge of how the release of enzymes in the joints of the body contributes to acute gout and rheumatoid arthritis. The chain consists of a heterogeneous collection of apparently unrelated facts, beginning with the discovery by the Russian Élie Metchnikoff in 1893 that a rose thorn, pushed into a starfish larva, causes cells to accumulate around the thorn. Following this came seven other unrelated contributions, by scientists scattered all over the world. They include a finding in 1931 by an American zoologist, Warren Lewis, that animal cells cultivated artificially in the laboratory engulf large globules of fluid from the surrounding nutrient medium, and one by Cambridge scientists in 1969 that sugar alters the inner structure of cells growing in laboratory glassware. Another vital but entirely separate discovery came from Lewis Thomas in 1956, working in New York on methods of dissolving clots in the bloodstream. In the course of his work he found that the enzyme, papain, injected into rabbits, caused their ears to droop.

Although biologists can now see pattern and significance in this apparently disjointed story,[3] it was not planned that way. The pieces can be put together only by hindsight. In other cases, one can discern an overall pattern in the process of discovery. This pattern, which recurs time and time again in all branches of science, has been perceptively described by the historian of science Thomas Kuhn in his book *The Structure of Scientific Revolutions*.[4] In the very earliest phase of the process, before scientists really know anything about a particular subject, all is guesswork and mystery. The field is open for even the most mystical speculation. Then comes the phase of discovery, precipitated perhaps by a practical development in experimental equipment or by a pioneering thinker. Almost inevitably, the new discovery conflicts with established ideas. The stage is set for revolution. Numerous theories are hatched to explain the new findings or observations, and the field attracts more scientists, drawn by the novelty and challenge of the newly discovered phenomena.

Next comes the crucial breakthrough, when someone makes real progress in suggesting a general pattern of explanation. The ideological framework he puts forward – his conceptual model of how things work – provides a basis for others in the same field to make substantial advances. Now the new field becomes fashionable, and in all probability its practitioners face heavy criticism from competitors who either cling to out-dated ideas or who see the writing on the wall for their own rival theories. Finally, the remaining gaps are plugged, and the jigsaw is completed. In this, the classical phase, what was revolutionary hardens into the new conventional wisdom. Kuhn uses the term 'normal science' for research done between these great revolutions – 'research firmly based upon one or more past scientific achievements, achievements that some particular scientific community acknowledges for a time as supplying the foundation for its further practice'.

In this pattern of events, even more clearly than in the capricious and anarchical chain of discovery illustrated above, we see science as a social activity. In Ziman's view, this is science as public knowledge at its most manifest, because 'The climate of professional opinion at any one moment is as important as the genius of individuals in determining the intellectual history of the subject.' A good illustration of this, and of the role of fashion in science, concerns the story of the positron.[5] The positron is a fundamental particle of matter. It is exactly like another such particle, the electron, except that it has a positive rather than negative electric charge. Like the electron, it reveals its presence by tracks as seen in an apparatus called the cloud-chamber. For many years, however, from around 1926, when they were first observed, to 1933, scientists either ignored these tracks or explained them away as 'dirt'. Then the American physicist Carl Anderson and the Englishman Patrick (now Lord) Blackett independently grasped their real significance. At that point, physicists realised that the positrons now actually observed were the very particles whose existence Paul Dirac had predicted in 1932 in an esoteric and unusually difficult theoretical paper. For about seven years of work in this field, however, scientists had ignored the evidence of their eyes because they had been taught, and had believed, that electrons were always negative, and that positive charges were always carried by the (much larger) nucleus of the atom. No-one had proved this. It was simply a fashionable belief, strong enough to influence profoundly otherwise excellent and critical scientists.

In his *Scientific Autobiography*,[6] the physicist Max Planck went as far as to say that 'a new scientific truth does not triumph by convincing its opponents and making them see the light, but rather because its opponents eventually die, and a new generation grows up that is familiar with it'. That is an extreme view, but it emphasises one of Kuhn's main points, the rigid orthodoxy of the scientific community and its staunch reluctance to consider really radical new ideas. Yet, as Dr M. J. Mulkay, an assistant director of research in the engineering department at Cambridge, argued in a recent book,[7] the very processes of social change which can maintain intellectual conformity in the scientific community also generate conditions favouring innovation. As a result, significant innovation happens mainly at the top and bottom of the hierarchy of science, not in the middle.

By definition, Kuhn's 'normal science' cannot continue indefinitely, for, as Mulkay explains, 'the problems become less and less significant and the professional recognition awarded those who provide correct solutions becomes less adequate'. As a result, researchers become less committed to the existing framework and more willing to rebel and formulate unconventional hypotheses. They are also more likely to migrate into other areas, where they can exercise their skills on meatier problems. 'In many instances this process of intellectual migration promotes a radical redefinition of an existing area of study, often in the face of strong resistance from those already engaged in the field. Just as often, however, intellectual migrants move into those new areas of ignorance which develop unpredictably on all research fronts. This second type of migration leads to the formation of a new social network and the gradual construction of a set of cognitive and technical standards defining the new area of investigation.'

But who *are* the rebels? They come from two principal quarters. Those at the very top of the profession have little to gain from conformity and some will therefore be willing to pursue heterodox ideas (though rarely to the extent of impairing a reputation built on past work within the conventional framework). Secondly, the very young scientist can afford to risk unorthodoxy. It may pay off, and if not many of his elders will forgive the impetuousness of youth. Despite the brain-washing rigidity of science education, the young scientist is also less likely to be firmly fixed in his views and more willing to entertain wild hypotheses. The one group from whom one should not expect innovation are those Mulkay calls 'persons of established middle status in science'. They cannot risk failure, and must carry on ploughing the same furrow in which they have already invested considerable time and skill. There are, of course, many of them around.

The crucial breakthroughs of science often hinge on contributions from people with very different characters whose ideas and approaches nonetheless complement each other. Such was the case with one of the greatest successes of biology in recent years, the unravelling of the structure of DNA and some crucial work that preceded it. James Watson's *The Double Helix*[8] tells his version of the events that culminated in the award of a Nobel Prize to Watson, Francis Crick and Maurice Wilkins for determining the structure of DNA (the hereditary material in the nucleus of living cells which carries genetic instructions from cell to cell when they divide and from generation to generation at fertilisation). Watson's book has been heavily criticised as a subjective account – just one man's version of the background to a historic piece of work and the personalities involved. It is bound to be such, and much of this criticism is misdirected. Be that as it may, when the events of 1952–3 are examined in historical perspective, it seems that considerable credit should also go to the earlier work of another scientist, the shy, aloof, uncommunicative Fred Griffith. Although, as Sir Peter Medawar has pointed out,[9] 'No Fred, no Jim', Watson's book includes not even a footnote to Griffith, who has been largely neglected in popular accounts of the conquest of DNA.

Griffith is important as prime mover of the train of events which led Watson and Crick to their momentous work in Cambridge because he first demonstrated a phenomenon called transformation. This occurs when a chemical (now known to be DNA) from one microbe causes a permanent change in the behaviour of another strain. Griffith reported this in 1928, following experiments on extracts from bacteria of the sort which then commonly caused acute pneumonia. However, as an infinitely cautious man, Griffith did not allow himself to speculate in his report as to whether it was DNA in his extracts that transformed the bacteria, though he discussed the possibility with his laboratory colleagues.

Paradoxically, it was Griffith's own scepticism and resistance to the full implications of what he had done which made his own work so valuable. DNA had been discovered in 1869, and some speculators had toyed with the idea that it might be involved in inheritance. But the climate into which Griffith's discovery was born was dominated by the belief that the carriers of coded instructions in living cells were proteins. There were many good grounds for this belief. In particular, much evidence suggested that DNA was far too small a molecule to carry all the information necessary for a role in heredity.

Into this climate of brash certainty came Fred Griffith, a modest man who even hated going to scientific meetings. He was also reluctant to write research papers, preferring to work away industriously by himself, with his characteristically meticulous technique. That is why his discovery of transformation – going very much against his own long-lasting belief that the properties of bacteria do not change in this way – commanded great interest and respect. Only because Griffith had done his work so carefully, with beautifully designed control experiments, did he allow himself to concede that 'there seems to be no alternative to the hypothesis of transformation of type'. So it was that his paper was not ignored – as would certainly have happened at that time with a less respected researcher. There was some scepticism among other bacteriologists, but no-one dared challenge publicly the reputation of one with such an impeccable record for scientific integrity.

In this way, Griffith, who pursued his crucial work on transformation no further, came to play an unparalleled role in modern biology. The train of events that was to lead to the Nobel prize for Watson, Crick and Wilkins soon passed to the Rockefeller Institute in New York. There, in 1944, Oswald Avery and his colleagues confirmed that DNA was the chemical which caused transformation. Despite some residual belief in proteins as the carriers of hereditary instructions, the race to spell out the structure of DNA, and show how it fulfils its role, was now on.

There could be few more contrasting personalities than the quiet, scrupulous, tediously careful Griffith, and Watson and Crick at Cambridge, extroverts who were prepared to back their brilliant insight. 'Both types of personality have their drawbacks and virtues,' writes Professor Martin Pollock,[10] 'but in this instance what seems to have been of really crucial importance was that they followed each other in the right order: a matter of natural selection, of course, but still worth remembering as an occurrence of critical significance.'

Science as a social activity

Science, then, is meaningless unless we consider the *community* of scientists.[11] Every individual research worker is a member of several different social groups, each intersecting with others. These include his immediate experimental team, the international community of people working in the same field, and the wider community of scientists in different disciplines who nonetheless complement each other in compiling a conceptual picture of the world. In kinship and mutual responsibility, between scientists throughout the world and scientists at different periods in history, the analogy that comes closest to describing the ideal is that of the 'community of saints' of the Christian church. Back-biting and politicking between scientists may make such an analogy laughable, but that is what scientists aim for and believe in, certainly in relation to the tactics and strategy of their trade.

The two principal formal means used in mediating this brotherhood are journals and meetings. Each provides a forum in which scientists can announce their results, fly kites, receive criticism and criticise the work of others, and nudge the evolutionary process of science forward.

The scientific journal is by far the more peculiar of the two channels of communication, and it has spawned several unhealthy tendencies which are now part of the very fabric of science. Time was when scientists used to publish books, usually small monographs, setting out their work in rounded form. Many, like Charles Darwin's *The Origin of Species*, published in 1859, were both learned treatises for fellow scientists and popular works at the same time. But just after 1660, the first scientific societies as we now know them were founded (England's Royal Society, for example, p. 68), and they began to publish periodicals, so that scientists found themselves writing papers as well as books. More recently, with the growth of science as an organised profession, scientific paper writing has become a breathless pursuit intimately tied up with a scientist's career and prestige. 'We may,' Professor Derek de Solla Price assures us, 'define a man's solidness as the logarithm of his life's score of papers.'[12] Be that as it may, there is little doubt that scientists publish papers as much for recognition and self re-assurance as to communicate their work to others. In any case, scientists in a particular field keep in touch with each other personally, by post etc., and seldom learn of discoveries for the first time by reading the journals. One study of physicists in Britain showed that researchers seldom read the journals but were nonetheless anxious to publish in them.[13]

What, then, is a scientific paper? Most begin with a section called the 'introduction', in which the author describes his area of interest and re-capitulates previous work in that field, indicating the specific problem about to be tackled. Next come details of methods used in his experiments and/or in calculating his results. These results appear in the third section, which must contain nothing more than the barest figures or findings, with no interpretation. Finally, under the 'discussion' section, the author can speculate and assess the significance of his results, drawing conclusions from them and interpreting them in the light of other scientists' research.

Unfortunately, this format gives a totally misleading idea of the process of scientific discovery. 'The scientific paper in its orthodox form does embody a totally mistaken conception, even a travesty, of the nature of scientific thought,' writes Medawar.[14] This is so because 'you have to pretend that your mind is, so to speak, a virgin receptacle, an empty vessel, for information which floods into it from the external world for no reason which you yourself have revealed . . . and in the discussion you adopt the ludicrous pretence of asking yourself if the information you have collected actually means anything.' In short, the scientific paper is based on the idea of induction, and does not reflect the various ways in which research really happens, which we discussed in the last chapter. It induces people to conceal mistakes which led to unexpected results, to fabricate plausible premeditated reasons for having done experiments which were productive but which happened by accident. This is both dishonest and misleading, not least for the student and would-be scientist. 'This very clear, ordered presentation of scientific discoveries produced a terrible inferiority complex in me when I was a student,' said the late Dr Hans Kronberger, a British nuclear scientist, some years ago.[15] 'I realised that I could never find things out in this way. In fact,

I was very worried whether I would ever make a scientist, or whether I would ever find anything out.'

Nonetheless, this is how papers are written at the moment. And this very artificiality is all the worse when you consider that papers are no longer simply communications to other scientists, plus a means of registering priority in making a discovery. Scientific paper writing has now become so important in advancing a scientist's career that it is virtually an end in itself. Even those who will not go so far as De Solla Price in quantifying a scientist by his score of papers, will still concede that the number of publications to a man's credit is vitally important to advancement in his career. This is one reason for the astronomical increase in the number of journals published in recent years and for such abuses as multiple publication of the same data in different journals. Widespread concern about this great tide of paper reached a crescendo during the mid 1960s. But times change. We have now grown accustomed to it.

More than that, an entirely new discipline has grown up, designed to estimate a scientist's worth purely by the impact he makes on 'the literature'. The idea is simple. Because important, seminal ideas and discoveries in science catalyse continuing waves of secondary, tertiary and further publications, one can construct a pecking order for the scientific community by listing the number of occasions when particular papers by particular individuals are cited in papers by others. There are now several publications, parasitic on the scientific journals, which specialise in providing such information. The only trouble is that the counting game cannot reflect even roughly the quality, as opposed to quantity, of research publications. For example, on a quantitative basis, the supremely important contribution to biological science in the past 25 years was a paper published in 1951 by Dr O. H. Lowry and his colleagues[16] on a new and sensitive method of analysing protein. It was a neat and useful technique, which was taken up immediately by other biologists and is widely employed today. Yet Dr Lowry himself would scarcely claim that it was the material of which intellectual edifices are made. Techniques are important in science, but this one has not of itself brought radically new understanding.

The high priests of citation worship are undismayed by the prospect of research workers lobbying to boost their quotation ratings, and of commercial interests that might be turned to this purpose. They ignore the insidious influence of fashion in science, and the impossibility of comparing meaningfully the frequencies of citation in different disciplines. They are unworried by the possibility of hasty judgments from citation counters, and forget that some of the greatest developments in science appear initially as heresy, left severely alone for a time by experts possessing the conventional wisdom. Above all, they do not seem to have a glimmer of the subtle ways in which scientific ideas are propagated, or of the dangers of prostitution inherent in any attempt to impose their paper doctrine on the scientific profession. The best comment on this sort of thing is that old story about the Crucifixion: 'I hear he was a great teacher,' said one of the centurions standing by. 'Yeah,' came the reply, 'but he never published anything.'

The proliferation of scientific journals and papers is a serious matter, but in one respect it can be exaggerated; it is not inherent in science itself. Many articles appeared during the 1960s arguing that the volume of published research findings was increasing at such a rate (doubling every seven to ten years, depending on subject) that experimental science would soon come grinding to a halt. Scientists would find it impossible to keep in touch with the vast and increasing amount of information which they ought to monitor.[17] In fact, this increase must be seen in relation to the evolution of science itself, which is accompanied by the appearance of new disciplines. In the early period of such a new discipline, there *is* a profusion of facts and ideas, but in time this is ordered into general principles. 'The ballast of factual information, so far from being just about to sink us, is growing daily less. The factual burden of a science varies inversely with its degree of maturity. As a science advances, particular facts are comprehended within, and therefore in a sense annihilated by, general statements of steadily increasing explanatory power and compass – whereupon the facts no longer need to be known explicitly, i.e. spelled out and kept in mind.'[18] Paper profusion is due more to the abuses stemming from careerism, which can be combated, than to the intrinsic nature of science, which cannot.

With more attention paid to the scientific literature in recent years, we also now know that research is duplicated *unintentionally* much more frequently than was thought before. The first systematic attempt to assess the size of this problem, made in Britain in 1964 by the Aslib Research Division,[19] suggested that more than 20 per cent of research work was unintentionally repeated by other research workers. Further surveys in the United States and elsewhere have revealed similar figures – which do, of course, imply a massive financial wastage. In one case an American electrical engineering company spent several million dollars and two years of the research time of a specially selected team in tackling a problem which had already been solved in Holland.[20] Information is expensive, but lack of information can be much more so. As Lord Rayleigh pointed out in 1884, rediscovering something in the library is often a more difficult and uncertain process than the first discovery in the laboratory.

Nonetheless, and despite the professional kudos attached to the great paper chase, it is not true that the race for priority is more phrenetic than in leisurely bygone days. Quite the reverse. Newton, for example, might be thought a likely archetype for the leisurely gentleman scientist – he worked before the professional term 'scientist' was even invented. Yet he was passionately concerned about establishing priority when engaged in his classical studies in mathematics and physics. He became obsessed with the task of ensuring the lustre and fame owing to him, and assembled a corps of young mathematicians and astronomers to help in 'the energetic building of his fame' (as the historian Frank Manuel puts it). The question of whether Newton or Leibniz invented the calculus created tremendous heat, and Newton's manuscripts contain at least ten different versions of a defence of his own claim to priority. Eventually, after Newton became president of the Royal Society, he set up a committee to evaluate the rival claims, packed the committee with his own disciples, and wrote an anonymous preface for the second published report on the controversy.

Nowadays, as shown by a survey by Professor Robert Merton and Elinor Barber at Columbia University,[21] scientists seem if anything less anxious about priority. This is probably because they are more aware than ever of the possibility that their own discoveries can be duplicated unknowingly by others. Thus scientists today are less likely than their forebears to assume that duplicate discoveries must be borrowed or stolen ones. Studying a cross section of 264 multiple discoveries, Merton and Barber found a successive decline in the frequency with which multiple discoveries have led to intense priority conflicts. Of the 36 multiples before 1700 in the sample, 92 per cent were contested strenuously. The figure dropped to 72 per cent for the eighteenth century, remained at about the same level in the first half of the nineteenth century, and declined to 59 per cent in the second half. For multiple discoveries recorded in the first half of this century the percentage was 33 per cent.

So the claim that priority conflicts arise from the hectic pace of modern science cannot be sustained. Scientists seem to have become *more* civilised about these things. As for the competitive edge, there is scant evidence that this is harmful in itself. On the contrary, there are many examples of scientists spurred on to more intense effort in the knowledge that others were on the same track. The race to solve the structure of DNA was just such a case, with the Cambridge team anxiously watching Professor Linus Pauling and his colleagues who were working on the same problem in the United States.

There is, however, a risk that a competitive climate may entice a scientist to try to publish work that is incomplete, badly presented, or hastily analysed – or all three – in an attempt to steal a march on a more prudent colleague. Such an individual can always place his paper in some journal, somewhere, because the standards of different publications vary so enormously. There are excellent journals publishing important and well-reasoned papers. Other journals carry mediocre papers, replete with shoddy work or sloppy reasoning, which simply clog up the system. There are also totally unnecessary collections of papers given at scientific meetings, the contents of which have appeared elsewhere, but whose repeat showing justifies the author's travelling expenses and lengthens his bibliography. Formerly almost all journals were published by the learned societies, but today commercial publishing houses are heavily involved. If anything, rather than benefiting from the purging effect which the commercial edge might have been expected to bring, the situation has worsened. Many a journal of dubious quality survives only because a sufficient number of research workers find it, reluctantly, an essential journal to consult for the occasional worthwhile paper. Usually, they avoid paying for such journals themselves by persuading their department library to subscribe.

Editors should, of course, reject incompletely presented manuscripts – which brings us to the role of the referee in journal publishing. Virtually the lynchpin about which the entire business of science revolves, he is the person to whom the editor of a journal sends a paper submitted for publication. Even a highly specialised journal will have several referees, looking after papers in particular areas within the journal's purview. The referee's job (for which he is not paid) is to say whether a paper referred to him presents worthwhile data and whether the arguments and methods are sound – in short, whether the paper is suitable for publication. He may recommend acceptance, outright rejection, or a middle course in which the author is asked to tighten up an argument or do more experiments.

Leaving aside totally inadequate refereeing – which, as with the few journals not using referees at all, can lead to the publication of erroneous and even bogus work – the commonest weakness of refereeing is in letting through incomplete papers. Take one recent example. In 1964 Professor John Yudkin and his colleagues at Queen Elizabeth College, London, first related atherosclerotic heart disease to sucrose in the diet. This stimulated a stream of papers seeking to confirm or repudiate the relationship. Other research workers have tried to prove that atherosclerosis is linked rather to consumption of animal fat, and that replacing this by vegetable oils reduces the risk of heart disease. Whoever is right, the situation is certainly complicated, and research workers try to conduct their studies in such a way that all possible dietary constituents are taken into account. Yet a paper appeared in an international medical journal in 1968 which set out to throw light on the situation but which omitted vitally important information about diet.[22]

In his paper, Professor Oglesby Paul reviewed results from a long-term study and claimed to establish that, though individuals who developed serious heart disease had consumed more sucrose than a control group, the difference was much less than previously reported, and was not statistically significant. However, Yudkin and his team had shown that older people take less sucrose than younger, the reverse being true for coffee and tea intake. Smoking also falls off with age. Thus any investigator seeking to assess the importance of these various factors in the genesis of heart disease should compare groups matched for age. Professor Paul did not record having done so in his survey. His paper did not mention the age of either the victims of heart disease or the corresponding individuals without heart disease. It seemed clear, in fact, that they were not matched for age, because Paul compared data for 66 men with heart disease with that for 85 'control' men selected at random.

The main reason why referees are often reluctant to exert their full influence and advise an editor to reject a paper, or insist that an author modifies it, is because they remember famous occasions when outstandingly important scripts have been turned down. The journal *Nature*, for example, rejected the first paper by Hans Krebs on what became known as the Krebs cycle – the centrepiece of living processes in the cell, and one of the foundation stones of modern biochemistry. It was turned down ostensibly on grounds of lack of space. Such spectacular errors of judgment must haunt even the most fastidious and fair-minded referee. Nonetheless, the critical side of scientific publishing could and should be greatly strengthened.

One snag about the refereeing system is that the referee is inevitably a colleague and/or competitor of the authors of papers submitted to him. Being human, and competing in the same event as the authors whose papers he referees, it must be difficult to play the hypothetical role of the totally disinterested, objective wizard. Professor Douglas Wright of Melbourne University has argued, on this basis, that referees' reports should not be anonymous as at present.[23] Wright cites instances of scheming referees, and points out that a referee can use his influence to recommend rejection and exploit the delay in publication to pursue the work himself. The first step in reforming the system, he suggests, is for referees to be required to provide a signed copy of their report to the authors. 'Why should the wish to publish a scientific paper expose one to an assassin more completely protected than members of the infamous society, the Mafia?'

This is an exaggerated comment, and no doubt most referees do their jobs conscientiously. Nonetheless, it *is* strange that, at the very heart of the open, critical structure of science we should find the anonymity of referees. They could continue to work punctiliously if their activities were not veiled in secrecy, with the added advantage of being able to discuss any disagreements openly with authors. As things are, the system is built upon a fallacy, that every referee is totally objective and has no interest beyond serving science. One important step towards improving the situation was taken in July, 1972, when *Nature*[24] published a detailed referee's critique alongside a paper on the chemical basis of learning in rats.

More open scientific debate is to be found at the scientific meeting. These range from the small colloquium or seminar, attended by a dozen or so people to discuss the state of knowledge in a particular field, to the gigantic international congress, usually held at a lavish resort and patronised by hundreds or even thousands of scientists. In terms of the formal agenda, the usefulness of a meeting is inversely related to its size. A hard day's talk and argument in a seminar room may be highly effective in resolving problems and clarifying knowledge in a specialised area of science. The International Congress, crammed with many simultaneous sessions, a hectic social programme, and lots of official lunches and banquets, usually achieves little in formal terms. Its chief merit is the opportunity offered to scientists from far-flung laboratories to meet each other informally, outside the lecture room, and establish useful contacts that will be continued afterwards through correspondence.

What is common to both types of meeting is the paper-reading session. A succession of speakers get up and present their research results in talks lasting about half an hour. These are usually less rigid in form than the scientific paper, and speakers will even occasionally confess how they *really* made a particular discovery – by mixing the wrong chemicals, for example, or forgetting an important stage in a manipulation. Each talk is followed by discussion and criticism from the floor. This works less usefully with a session at a mammoth congress because the participants soon become punch-drunk with the sheer weight of the proceedings, often lasting for a whole week. Moreover, with a huge gathering the few scientists able to contribute useful comments on any one paper tend to be heavily diluted by people who are experts in an adjacent field but who feel apprehensive about trespassing on to unfamiliar territory.

The International Congress does, however, offer compensations.

The title, list of invited speakers, and location often sound so impressive that a scientist can persuade the department where he works to pay his travelling expenses. If he is going 'by invitation', then so much the better. Every congress usually sets aside one or two occasions for 'starred' lectures by celebrities, and it is considered a great honour to give one of these.

Ethics and values

One crucial consequence of the fact that science is a social activity, is the demolition of the legendary belief that it is an entirely amoral, intellectual game of problem-solving. On this view, scientists are desiccated calculating machines, whose work is determined solely by meter readings and what they can deduce from them. The truth is very different. Far from being amoral and coldly logical, science actually generates values. These include intellectual humility, an unusually acute regard for honesty, respect for the revolutionary and the apparent crank, and stress on the importance of co-operation. These are not optional extras for the scientist; they arise directly out of the pursuit of science. The degree to which a scientist lives by them will be reflected in the health of the scientific community and in an individual scientist's long-term success in his trade. Even the most inveterate liar must, if he is to succeed in science, cultivate a deep respect for the truth when he is about his work.

Of course, as we have seen, there *are* opportunities for chicanery and cheating. But science simply cannot progress unless honesty and humility before the facts are cultivated and adhered to by its practitioners. It may take many years for dishonesty to be detected, as with the hoax of the Piltdown Man.[25] Other such abuses will be spotted quickly and demolished by open criticism and cross-checking. Always, the inexorable, self-correcting movement will purge science of falsehood. The distinguished American philosopher of science James Conant even suggests[26] that the traditions and conditions of research are such as to make rigorous performance axiomatic: 'Would it be too much to say,' he asks, 'that in the natural sciences today the given social environment has made it easy for even an emotionally unstable person to be exact and impartial in the laboratory? The traditions he inherits, his instruments, the high degree of specialisation, the crowd of witnesses that surrounds him, so to speak (if he publishes his results) – these all exert pressures that make impartiality on matters of his science almost automatic. Let him deviate from the rigorous role of impartial experimenter or observer at his peril; he knows all too well what a fool so-and-so made of himself by blindly sticking to a set of observations or a theory now clearly recognised to be in error.'

That is an extreme view of the degree to which the social climate forces scientists to embrace the highest ethical standards in their work. But our reluctance to believe something like it comes largely from observations of scientists outside the laboratory. Clearly, they by no means always take their scrupulous honesty and intellectual humility with them when they deal with other matters – even, as we shall see in Chapter 7, when involved in the politics and organisation of their own trade. Nonetheless, the ethical standards of the scientist about his business in the laboratory are required to be the very highest.

In our glance over the scientific community in this chapter, we have already seen hints of one of its less attractive features, which I shall discuss in greater detail later in the book – its autonomy. True, scientists belong to the wider community in which they live, and they reflect its beliefs and values. And someone has to pay them to do research. But in the day-to-day conduct of their affairs, scientists have much greater freedom and autonomy than other sections of the community. One unfortunate consequence of this is the creation of unstoppable bandwaggons in science, the continued motion of which becomes a matter of professional pride for those at the wheel. A laboratory or research institute devoted to a particular research project is much more than just bricks and mortar and a list of staff; the prestige and career prospects of the team are intimately tied up with the fulfilment of the research programme. This can be both dangerous and wasteful. As Professor Edward Shils writes,[27] 'The fact that a given section of the cosmos offers fascinating and important problems for a decade is no guarantee that it will continue to be as important twenty or thirty years later. Yet the multiplication of scientists who are not very creative but who could play a useful part in the conduct of "normal science" constitutes a claim for and a commitment to the future support of the kind of research which they can do and in which they have proved their competence. They become a vested interest in the republic of science and not necessarily one which is conducive to the common good of science or of the larger community.'

Despite the unexpected severity of its internal code of conduct, and to some extent because of this, science can all too easily be out of touch with the needs, goals and aspirations of the society that nourishes it.

Chapter 4
Learning the craft – and its rewards

Initiation into the scientific community is a formal process of qualification, in practice every bit as exclusive as entry to the law or priesthood. As I argued in the last chapter, science depends upon tolerance and a broad-minded readiness to listen to unorthodox ideas. But if they are to stand a good chance of being considered seriously by the established practitioners of science, such ideas must come from other established practitioners. There are still a few amateur scientists, without the recognised qualifications of the craft, who make acceptable contributions in such fields as astronomy and wildlife studies. But their numbers are insignificant in proportion to the international corps of 'real' scientists admitted to what Michael Polanyi calls the 'Republic of Science'.[1] And when an amateur leaves the modest, humdrum field where he is tolerated and proposes a controversial new theory, he has a considerable up-hill struggle against professional inertia. When such a person, without degrees or titles, does make good, it is a very special occasion. He will be welcomed at scientific meetings with effusiveness tinged with embarrassment, like a distinguished coloured politician greeted by white racists. Moreover, as science becomes more costly and dependent upon sophisticated equipment, the amateur has even less opportunity to pursue worthwhile original research.

The standard qualification obtained by most professional research scientists is the doctorate (Ph.D. or D. Phil.), taken after a 'first degree' in science. It was in Germany during the nineteenth century that the doctorate was first established as an award for achievement in research. Aspiring scientists from throughout the world flocked to Heidelberg, Göttingen and the other German universities to sit at the feet of the revered figures of science, learning the craft in true Socratic fashion. This was the origin, too, of research schools. Consisting of a Grand Old Man surrounded by his younger scions, each was dedicated to solving scientific problems in a particular area. The flourishing of such schools was to a large extent responsible for the dominance of Germany in the pure sciences during the last half of the nineteenth and the beginning of this century.

The German universities awarded their doctorates (D. Phil.) for original research work, written up in the form of a thesis – a very much longer document than a scientific paper, but following the same structure. The author assesses previous work, outlines the problem to be tackled, describes his methods and results, and discusses his findings and their significance. The same pattern was adopted in the United States, where the first doctoral awards were made in 1861 at Yale. By the 1920s, over thirty States in America had institutions awarding doctorates, and the Ph.D. established itself as the obligatory qualification for leadership in professional science. In Britain, it was not until after the First World War that the Ph.D. became the normal result of two or three years of successful scientific research.[2] Again, however, the doctoral system quickly emerged as the conventional means of producing professional research workers. Following the German pattern, British and American universities began to attract aspiring scientists from other parts of the world, so that doctoral training has greatly stimulated the international intercourse of science.

With the growth of science as an organised profession, has come the conveyor-belt system (though a surprisingly capricious one) of turning out scientists. It starts when a pupil at school shows promise in a science subject, and his teacher, mindful of the greater glory of the school, propels his young charge towards a university degree course in that subject. Then comes the strait-jacket of the science degree course itself. With one or two outstanding exceptions, most university science courses are designed to produce technocratic specialists – professional scientists. They are therefore heavily fact-ridden, and are strictly vocational, lacking the broader perspectives that would be helpful for both future research scientists and for those undergraduates who will later enter teaching or other employment outside of scientific research. As Professor Freddie Jevons observes, 'studying science becomes stenography plus memorisation'.[3]

The straitjacket tightens with the consummation of the first (bachelor's) degree. If he does well, the newly fledged science graduate will find increased pressures on him to go into research. It is not only that, even in times of austerity, his good first degree opens doors, widens his choice, and provides a further incentive and challenge. The whole flavour of his academic surroundings is such as to suggest that he will let the side down, and be less than fair to himself, if he does not pass on to the next logical stage. 'The most esteemed model of the scientist available for emulation is the academic scientist.'[4] Despite the obvious fact that we need more scientists in other walks of life, for many academics a first-rate science graduate who rejects the possibility of research is still looked upon as a loss to the system that has nurtured his talents, if not a total failure. So it is that, even today, many people who have never made a conscious choice of a career in research find themselves embarking upon one.

The young graduate may stay on to do his postgraduate work in the same department where he attended lectures as an undergraduate. Alternatively, through reading an advertisement or advice from friends or teachers, he may hear of a research studentship in a department elsewhere in the country or abroad. Unless he has private funds, the student must obtain one of the research grants that are awarded each year by government bodies such as the Science Research Council in Britain, by the Foundations, and by private industry. There are peculiar anomalies at this stage, some award-giving bodies insisting on a certain standard of first degree, others asking no questions. Also, many such bodies give an award for only a fixed period. In Britain, this is often three years – the average time which it takes to gain a doctorate in the UK – but it is usually longer in the USA. Others have a more open-ended arrangement and are willing to extend a grant, subject to satisfactory progress. It is also possible for a graduate to prepare for a Ph.D. degree while working as a member of the staff of a university department or allied institution and earning a salary. In such cases, the qualifying period is usually longer. At present only a minority of doctorates are gained in this way, though in the USA many Ph.D. candidates spend a substantial time demonstrating in the laboratory and teaching in exchange for their grant.

Once over his preliminary hurdles, the student begins work on a research project, guided by his supervisor, who is usually himself a Ph.D. of some years' standing. Regulations vary from one university to another, but typically the form of words requires that 'a student shall engage in full-time research for a period of nine terms. Afterwards he must present the results of his work in a thesis, and undergo an oral examination in which he should display evidence of his thorough understanding of the work and of his ability to design and carry out original experiments.' Both the oral examination and the evaluation of the thesis are in the hands of the student's supervisor and an 'external examiner' – an expert in the relevant field of research, who has been appointed on the recommendation of the supervisor and is usually a friend of his. Increasingly, and particularly in graduate schools in the United States, a postgraduate student doing research must also attend lecture courses and/or seminars. These are similar to, though more advanced than, those he has experienced as an undergraduate. Some universities have established a master's degree (M.Sc. or M.Phil.) as an intermediate stage towards the doctorate. Some make this obligatory for graduates with anything other than a first class first degree. Again regulations vary widely, and the M.Sc. may be gained for either a short original research project similar to Ph.D. work, for an examination based upon a lecture course, or both. In general, in Britain a M.Sc. signifies a higher degree based on course work, a M.Phil. one based on research. In each case, however, the aim is to erect a further testing barrier so that only the cream of candidates are allowed to proceed to the doctorate.

The actual Ph.D. system itself is shot-through with anomalies and frustrations – which is itself an anomaly for a system that plays such a crucial role in determining the future careers of professional people so useful to the community. Neither a student nor a member of staff, the research student inhabits a no-man's land where he is never really sure of his status and where he is denied the advantages of belonging to either of the other two groups. To an alarming extent, his fate is entirely in the hands of one man – his supervisor – whose personal quirks can affect profoundly the young man's future career. In particular, the way the research student is guided in his work varies enormously between one university department and another.

One supervisor will interpret Ph.D. training simply as an

introduction to research methods – as a means of learning how to plan and carry out experiments and interpret the results. In contrast to the rote-learning of an undergraduate course, the student acquires these skills by investigating, under close supervision, a particular scientific problem. If, at the end of the agreed period of, say, three years, the student has developed the necessary qualities and abilities, and can discuss his work intelligently, then he should get the degree – whether or not he has succeeded in solving the problem that was his original task.

A variation is the case where a student is blatantly exploited as a 'pair of hands', doing tedious and repetitive work which could well be left to unqualified laboratory assistants. He will be told what to do, with daily instructions from his boss, and will require little imagination or intellectual brilliance to satisfy the doctoral requirements. This happens typically in a very large department tackling a mammoth research programme that breaks down into many smaller projects, each differing only marginally one from another. An example is the department devoted to studying the chemical structure of parts of certain plants or microbes. In practical terms, such a research programme requires many hundreds of thousands of analyses of material from different related organisms. When all the analyses are complete, they will provide insight into the evolution and behaviour of the organisms. But meanwhile, someone has to tackle the drudgery of the benchwork. One way is to have droves of research students – probably all supervised by the same senior scientist – each of whom analyses a small number of samples of material. They all use the same analytical techniques. The theses they write contain virtually identical sections on methods, background, and discussion. Only the details of the structure of one student's chemical samples differ from those of his colleague on the next bench.

Nearer to the German origins of the doctorate is the more leisurely case where the student becomes apprenticed to a great man, works under his tutelage for some years, and hopes that a little of the greatness rubs off. One still encounters students like the hapless Albert Woods in William Cooper's novel[5] who, when he turned up for work only a week after being warmly welcomed by his professor, found that he had been totally forgotten and could not find bench space on which to work. Such students tend to receive brilliant guidance, but to receive it infrequently. They are handed a research problem and are expected to produce a solution – whether it takes the regulation time, or less, or a good deal longer. The result is that the weak flounder, and the clever thrive. The supervisor – perhaps the only expert in the country in his own field – may disappear abroad for long periods, leaving the student to push on alone as best he can.

Another anomaly concerns the publishing of research results while in the apprentice stage. Should research students publish their work in the journals and present papers at scientific meetings? Some supervisors consider that learning to communicate in this way is an important part of the training of a future scientist. Others believe that a research student should confine himself to writing his thesis until he has been admitted to the magic circle. Again, one supervisor will direct his students in writing their theses, to the extent that the finished products are really his work, rather than those of the students, and contain his very own ideas and interpretations. Another will argue that because the supervisor is also one of the two examiners who assess a thesis, it should be entirely the work of the student. Finally, the Ph.D. oral examination can be anything from a stiff grilling to a mere formality. A student with an acceptable thesis can still fail on his viva. Many escape with a chat over a pint of beer.

These are serious discrepancies, which obscure the value that can be placed upon the doctorate. National differences are less important. Everyone knows that newly-qualified American Ph.D.s are more mature than those in Britain, because their training is significantly longer and is carried to a more advanced level. Compared with the three year average in the UK, the average time in the USA is six years.[6] The differences *within* the two countries are less acceptable.

The student-supervisor relationship is a mutually beneficial one, like the biological phenomenon of symbiosis. The student learns his trade and the supervisor extends his own research activities and reputation. Just as the number of research papers to his credit is an index of a scientist's professional achievement, so possession of a clutch of Ph.D. students is an outward and visible sign of prestige. A supervisor's name also appears alongside that of his students in all research publications describing their work. With a particularly bright student and an inept supervisor, the relationship can easily become parasitic, and there are many stories in the scientific world of unscrupulous supervisors who took more than a fair share of the credit for their students' discoveries. In recent years, with the growth of contract research, pursued in universities on behalf of industrial sponsors, a further potential abuse has appeared. This is the use of research students to do work which both promotes a supervisor's personal and financial standing, and is designed to boost the profits of a private company by solving its technical problems.

It is not surprising that there are periodic outcries from the research student fraternity. Such protests would be more shrill but for two reasons. The peculiar relationship between master and apprentice, which all too often smacks of exploitation, makes it risky for the aggrieved student to say publicly what he feels privately about his lot. At the very outset of a career in a profession where status and intellectual reputation are of prime importance, it would be foolhardy to incur the displeasure of one's seniors. What the research student most desires is to gain his doctorate and get out. Moreover, research students are not well organised as a political group. There is no effective body or corporate mouthpiece through which they can lobby for change, partly because most are far too busy for such activities. In Britain's National Union of Students, for example, their presence and their concerns are submerged by those of the much larger membership of undergraduates and other students taking first qualifications.

A classic historical example of exploitation by a supervisor is to be found in Edward Lurie's biography of Louis Agassiz, the eminent nineteenth century naturalist.[7] Agassiz was a charismatic teacher, and many of his students became intensely devoted to him. Yet most of them eventually fell out with him, and one of his students at Harvard publicly accused him of dishonestly appropriating their work. As long as his students were loyal and accepted a totally subordinate status, Agassiz rewarded them, but he would not tolerate any aspirations towards independence. Other famous cases of unfairness in research supervision include Sigmund Freud and Karl Pearson, both autocrats in dealings with their students. Another was Humphrey Davy, who opposed the election of his most illustrious student, Michael Faraday, to the Royal Society, for reasons that are unclear but seem to have arisen out of their former master-and-student relationship.

Sourness and exploitation has, of course, much more opportunity to flourish in today's conditions, in which the Ph.D. is an obligatory qualification for the aspirant scientist. There is only one way of securing a doctorate, and both student and supervisor know it. Not surprisingly, a survey conducted a few years back by Bernard Berelson[8] showed considerable evidence of discontent. In a substantial study of recent American Ph.D. recipients in different disciplines, he found that 57 per cent of those qualifying in microbiology, biochemistry, or biophysics agreed that 'major professors often exploit doctoral candidates by keeping them as research assistants too long, by subordinating their interests to departmental or the professor's interests in research programmes etc.' Corresponding figures were 52 per cent for people qualifying in physics, 40 per cent for those in zoology and geology, and 37 per cent for psychologists, but only 28 per cent for mathematicians. These, of course, were all recently qualified Ph.D.s, free of any inhibiting restraints of their former life. At least the same percentages of current research students today would have justifiable cause to complain.

It would be wrong to argue that the Ph.D. system is corrupt. It can, and sometimes does, work superbly well – but not consistently so. As a means of access to a profession, the system is staggeringly capricious and ill-organised. It is a process which, like education generally, the more able people manage to survive rather than thrive upon.

The greatest indictment of science training, however, concerns not its methods but the narrowness of its content. From the earliest school lessons and textbooks onwards, the intellectual framework of science is presented as something both fundamental and wide-ranging in its universal significance, and yet isolated from other human concerns. This is the origin of the subtle arrogance of many scientists. They know that they hold the conceptual keys to understanding matter, energy, the universe, the processes of life. This bestows a philosophical affinity with their environment that is infinitely more important than mere economics, politics or artistic endeavour. Their training has both neglected these other areas and engendered a subconscious readiness to dismiss them as mere epiphenomena of transient interest.

That ghastly cliché about 'scientists playing God' partly originates here, and not with the power that science has put into Man's hands. Possessing a common esoteric language, and sharing the great intellectual edifice of science, scientists develop a mutual sympathy allied with superiority over outsiders every bit as insidious as that of a secret society or religious group. It is a powerful and seductive feeling.* Even the biologist who is an ignoramus in the realm of electronics, or the physicist unfamiliar with modern biology, knows that he possesses the keys to explore these other areas. Each feels at home with the conceptual framework of science – with molecules, subatomic structure, cells, forces, radiation, and the theories of cosmology. And each is familiar with the rigorous intellectual methods needed in investigating unfamiliar territory. The superiority engendered by no means always appears as overt haughtiness, and is to be found tucked away in the subconscious of many an apparently modest, humble scientist. It has something in common with the superiority of the arts man who feels that others are blind to the qualities of his particular art form. But it is all the more powerful because it deals, like religion, with eternal profundities.

There will be no great change in this regard until science teaching, in both schools and universities, is placed firmly in its social context. If you begin by telling a schoolboy about the eternity of matter and energy, and over the years add layer upon layer to the growing conceptual framework, without any reference to the interaction between science and society, then you will tend to produce naivety, xenophobia and intellectual arrogance. The consequences are numerous. One is the conspicuous failure of many scientists to think historically. 'The only thing wrong with scientists is that they don't understand science,' writes Eric Larrabee. 'They don't know where their own institutions came

* It can also appear as grotesque intolerance of the outsider, overriding the tolerance we discussed in Chapter 3. *The Velikovsky Affair* (edited by Alfred de Grazia, Sidgwick and Jackson, 1966) shows the viciousness of the scientific community – in not only attacking a scientist with a heterodox theory, but also trying to prevent him from publishing his views.

from, or what forces shaped and are still shaping them, and they are wedded to an anti-historical way of thinking which threatens to deter them from ever finding out.'[9] This static mode of thinking does not derive from the methodology of science in looking at frozen instants of reality. Change and evolution are at the very heart of both biology and the physical sciences. It comes rather from the failure of science teaching, from school onwards, to show how science reflects the society in which it arises, changes that society, and interacts with it (as we shall see in the next two chapters).

A result of the non-historical flavour of science teaching is that many scientists are politically inept and commercially innocent. Politicians suspect scientists, seeing them as insensitive and dangerous, while industrialists complain that even the brand-new Ph.D. is often so narrow-minded and naive that it takes several years before he is of much use to the company. To be sure, things are changing, but slowly and partially. There have been great improvements in school science teaching in the past ten years – pioneered, for example, in Britain by the Nuffield Foundation and by such projects as ChemStudy in the USA – but these have been more concerned with the conceptual content of teaching than with the social relations of science. And while some universities run courses on the history and philosophy of science, for science undergraduates and postgraduate research students, these tend to be take-it-or-leave-it courses grafted on to a science course rather than integrated with it. Broadly-based degree courses with such titles as 'liberal studies in science' are also increasing, though these are mostly designed to promote the flow of educated scientists into fields other than science. This is an important activity, but not one that will help future professional scientists. Over all, the flavour has changed surprisingly little since the last world war.

It is true, of course, that many technologists receive a somewhat broader training than that of the research scientist. Civil engineers, for example, learn some economics and management techniques, because these are essential to their future work. Building a new motorway is not simply an engineering exercise. Such a project is intrinsically complicated, requiring skills other than technical ones, and has to be planned with sensitive regard for its social consequences and acceptability. But in this book we are talking primarily about science, and even such self-interested breadth of training as engineers receive is lacking in the education of most of tomorrow's scientists. Yet, as we shall see, many of the most far-reaching developments nowadays are initiated and brought to practical fruition by scientists, not technologists. The contraceptive pill in recent years (and, in the future, a pill to arrest ageing) are two examples.

The job crisis in science

The narrow training of the scientist makes him appallingly vulnerable in times of economic recession, such as began to afflict both the United States and Britain in the late 1960s. In part, the great US recession in science started as a consequence of a general trade recession in the country, and indeed in the whole western world. Whenever industrial companies find their costs rising and their profit margins falling, they search around for economies, and where better to begin than in the research department? What could be more dispensable to short-term profitability than speculative, long-term science? Ignoring the awkward question of what to sell once the fat store of existing know-how is exhausted, there is no-one less likely to be missed than the research boffin. The sales manager, the advertisement staff, and the tea-lady are all much more important.

The other major contributory factor to the US situation was the cut-back in the space effort. In the early days of the Apollo programme, research teams in industry and the universities throughout America were caught up in its contract research work. So, when the space programme began to slow down in 1969–70, many specialists found themselves without work – and specialist research facilities lay idle.

Yet the American educational machine continued to churn out an ever increasing number of science doctorates. 'Despite gloomy reports from the job market, the Ph.D. pipeline is operating at peak production,' wrote Phyllis Lehmann in the middle of 1971. 'In fiscal 1970, 14·4 per cent more Ph.D.s were granted than in 1969 – an increase that represents the highest growth rate since the immediate post-World War 2 period.'[10] Paradoxically, an economic recession does not necessarily deter students from pursuing higher degrees. Some students may, in fact, decide to stay on for doctoral work, after taking a first degree, for the very reason that jobs are scarce. They hope that the climate will improve later. On the other hand, postgraduate students already in the midst of their research may graft all the harder, aiming to get out quickly before the situation worsens. This, of course, increases the output of Ph.D.s in the short term.

A similar situation has obtained in Britain, and has accentuated the misery of those science graduates turned out by university departments that are out of tune with the outside world. Such departments tick quietly over, processing their annual intake and turning out graduates by academic momentum. Again, the better ones survive, but particularly in times of recession the less talented graduates discover, too late, that their course has fitted them for nothing in general or in particular. Emerging with a third class or pass degree, they start casting around desperately for a job. Many turn to teaching, for the worst possible reason. The course they have taken was designed for the aspiring scientist, and lacked almost everything that really counts in the classroom. Such grotesque consequences could be avoided if universities were to adopt the type of proposals put forward by Professor Brian Pippard and his colleagues at the University of Cambridge.[11] They believe that science degree curricula should be redesigned on the basis of a broad two-year general course, followed by a further two years of specialised work, for those who would truly benefit, towards the M.Sc. degree. Such rethinking could resolve many problems of scientific manpower, job satisfaction, and educational resources. Pippard's proposal has encountered passionate criticism on the grounds that it entails an 'élitist' approach to education. This is anathema to those who believe that a country should make honours courses in biophysics, endocrinology, and any other speciality available for all those who request them, regardless of qualifications and aptitude. The Pippard system would, however, put a timely brake on wasteful over-production of specialists, while giving all science graduates a much needed breadth of training.

It is, of course, highly desirable that more scientifically qualified people move into occupations other than science, so lending their special skills to those areas. A university degree in science, higher or lower, should not be seen simply as a meal ticket in a particular research discipline. But as things are at present the vast majority of scientists who enter employment outside science do so because of a lack of the type of post they really want within science. The shake-out of scientific expertise has not occurred through conscious government or educational policy, or through individual choice. Much of the redeployment, too, in recent years, has been into occupations which in no sense whatever harness the special skills of the scientist. Even a qualified scientist who has to work in a research laboratory, but as a technician, doing other people's experiments rather than his own, underlines the failure of the system through which he qualified. He faces personal frustration, and the money invested in his training is lost to the community.

Apart from the shake-out of scientists into non-scientific jobs, versatility and adaptability are required increasingly within research itself. The progress of science depends more and more upon what is fashionably called interdisciplinary research – pursued by mixed teams of scientists trained in different branches of science, and by scientists who have moved across the traditional boundaries of science. But this has had little effect so far on the training itself, and graduates are often reluctant to change disciplines after qualifying. There is, however, little justification for denouncing a new graduate for having a narrow, vocational view when, all too often, his entire career has been built on the supposition that an honours degree or Ph.D. in chemistry *does* imply a career in chemistry research, and that anything different means failure. Least of all does this argument hold much sway with the nuclear physicist who is applying his intellectual training to the job of barman or petrol pump attendant. What is needed – and urgently – is that much more realistic advice about career prospects and purposes be given to school pupils and undergraduates beginning science courses, and that such training includes an assessment of the place of science in the world. Beyond that, as I shall argue in Chapter 11, we should begin to work out why we train scientists at all.

Rewards and honours

One *beneficial* result of science training as it is today is that few youngsters enter the profession in search of high financial rewards. They pursue science out of curiosity, in search of intellectual prestige, because they enjoy studying the subject, and for many other reasons. The promise of money is low on the list. This may well be changing. With the growth of industrial consultancy work among university staff – and such symbols as the American system of 'meritorious professors' (at 50,000 dollars or so a year) – the aspirant scientist now has earlier contact with the seductive influence of high financial rewards. But at the moment, the fame, power and influence which befall those who receive the highest honours of the scientific world are much more potent forces.

Of these honours, the most prestigious by far are the Nobel Prizes, endowed by Dr Alfred Nobel[12] at his death in 1896. Each year three are available in scientific disciplines – chemistry, physics, and medicine and physiology. There are interminable arguments about some of the recipients, but there is no doubt that, once a Nobel laureate has been given his accolade, his fellow scientists treat him with high reverence. This counts for considerably more than the cash (now over £35,000) that goes with the award. The selection process is a long and scrupulous one, vested in the Swedish Academy of Science in the case of the physics and chemistry prizes, and the Caroline Medical Institute in Stockholm for the prize in medicine and physiology. Like the Roman Catholic Church sifting and scrutinising candidates for canonisation, the selection committees are subject to many pressures in choosing the recipients of their unique accolade. Not the least of their problems is that of interpreting Nobel's instructions that the prizes should go to 'those who, during the preceding year, shall have conferred the greatest benefit on mankind'. In practice, it by no means follows that the greatest benefit on mankind is bestowed by the 'most important discovery' in each of the three fields – the other criterion Nobel laid down. Moreover, it is virtually impossible to award prizes for discoveries made in the previous year, because of the months or years taken for their significance to emerge. The time required for the work of the selection committees is also prodigious. There is widespread criticism of life-long delays before the earlier work of elderly scientists has been recognised by the Nobel award – though this also strengthens the certainty that such choices are wise, and thus heightens the prestige bestowed.

An interesting analysis of Nobel prizes carried out by Dr Harriet Zuckerman of Columbia University revealed a close network of pupil-teacher relationships.[13] In one case, seventeen laureates could be linked together in a single pupil-teacher family tree, covering five generations and starting with the German chemist Johann von Baeyer who won a prize in 1905. Dr Zuckerman also found that as many as half the laureates who had their doctoral training in America did so at only four universities – Harvard, Columbia, Berkeley, and Princeton. Discussing the effect of winning a Nobel prize, one physicist told Dr Zuckerman, 'Very few people have survived it with a whole skin.' Like a sudden win on the football pools, the zoom to stardom can have disrupting effects, even for a scientist who is already well-known and respected in his field. Productivity at once decreases – from an average of 6·2 papers in the five years before they won a prize to 4·2 in the five years afterwards, in the case of the scientists questioned by Dr Zuckerman. And though new Nobel laureates, secure with their fame, become more generous about giving greater prominence to the names of colleagues in jointly-written papers, barriers also tend to grow, and arguments about credits for the Nobel work itself often arise.*

Following the Nobel Prizes, election to a distinguished scientific society is next on the list of honours coveted by working scientists. Of these, Britain's Royal Society is unquestionably the greatest. Election to its fellowship (which bestows the magic initials FRS), or to foreign membership in the case of overseas scientists, means accession to the oldest and most exclusive scientific society in the world. In the seventeenth century, the Royal Society began as a forum for discussing scientific discoveries and was centred around

* Most recently in June, 1972, when the physicist Oreste Piccioni filed suit against two Nobel laureates, Emilio Segre and Owen Chamberlain, alleging that they cut him out of participation in an experiment which he had designed at the Lawrence Radiation Laboratory, for which they won a Nobel Prize in 1959.

men such as Newton, Halley, and Pepys. The Society received a royal charter from Charles II in 1660, and began its long, though chequered, history in which it symbolised scientific excellence and took a close interest in the social significance of science. In 1901, however, the fellows decided to exclude economics, philosophy and the other 'human sciences', which inevitably rendered its proceedings more specialised and esoteric than before. Today, the Royal Society elects an annual quota of 32 new fellows, now including the previously neglected categories of applied scientists, engineers, and chairmen of industrial companies. The exclusiveness and the honour of being one of the 700-odd fellows remain – the pinnacle being election as the society's president.

'The Royal Society,' writes Anthony Sampson in his *The New Anatomy of Britain*,[14] 'has grown up alongside monarchy, government and church over the last 300 years, and it is in constant danger of being lured into the sleepiest regions of the establishment. It can easily get absorbed in its own mumbo-jumbo, and it has been accused of having lost every faculty (like the Pacific Palolo Worm) except that of reproduction. . . . The Royal Society typifies the older scientists' self-portraiture as a priesthood with their own rituals and missions.' Of the many institutions of science, the Royal Society has been least affected by the various movements of critical science and anti-science which we shall consider later in this book. Safe in what Sampson calls its 'ancient citadel of self-admiration', the Society clearly feels little need to reckon with such transparently erroneous movements. Like a great church that refuses to adjust to a changing world, this loses it some support, particularly among the young. At the same time, it heightens its prestige. The Royal Society is a sitting duck for irreverent criticism, but election to its fellowship is highly prized by even its most vociferous critics. It has its imitators, in Britain and other countries, but they all lack the lustre of the Royal. Unlike the promoters of a new international golf tournament, who can boost the attraction of their event over ancient competitions by using vulgar cash, imitators of the Royal Society simply cannot compete with a body whose currency is unique honour and distinction, sanctioned by history.

The counterpart of the Royal Society at the apex of American science is the National Academy of Sciences. This was established during the American Civil War and is, like the Royal Society, a tradition-bound, self-perpetuating body. Election is by a highly complex screening process in which the existing membership (totalling about 780) select 45 new members each year. 'The Academy is supposed to encompass the most professionally distinguished members of the American scientific community,' writes Dan Greenberg, 'but with the average age of admission being 49·1 and the average age of the total members 61·6, there are many distinguished young scientists outside the Academy and, regardless of age, there are some rather undistinguished ones inside. On the whole, however, the Academy represents the best of post-middle-age science in America.'[15] As election to the Academy ranks next to winning a Nobel Prize for an American scientist, it is something of an embarrassment when a non-member receives his urgent telegram from Stockholm. Since 1950, the Nobel Prize has been awarded to nine American scientists who were not members of the Academy at the time of the award. In all but two cases, Academy membership was swiftly bestowed upon them.

Are scientists born or made?

In discussing science training, and its rewards, I have not considered directly the psychological question of why people become scientists. There is a mass of literature on the subject, largely mutually contradictory. Lawrence Kubie, for example, argues that the motive for entering science is often a need to gratify unconscious neurotic desires.[16] Liam Hudson believes that the type of person who becomes a scientist likes to answer questions where clear answers are to be found – in contrast to the arts man's preference for the open-ended question.[17] And so on. Such explanations are probably true for particular individuals. Today, as I argued earlier, sheer inertia plays a large part in propelling people on to the scientific tramlines, just as in earlier years they would have gone into the church. But we should not ignore the beguiling influence of prestige and honour. As John Ziman points out,[18] science is like any good game. 'It gives one the chance of showing off, and winning a round of applause, before the whistle blows.'

Chapter 5
Science and society

Contrary to what schoolboys are often taught, science is not an esoteric subject totally isolated from the other affairs of men. As well as its own internal logic, science is shaped by the personal beliefs, education, and political attitudes of its practitioners. The institutions of science, and the deployment of its practical results, reflect in part the history, power structures, and political climate of the supportive community. At the same time, science constantly changes the world in which we live, in many subtle ways. It would be difficult to exaggerate its Promethean power. The most obvious examples – nuclear weaponry in particular – require no emphasis. What does need to be highlighted is the fact that any one of the thousands upon thousands of new discoveries, products, processes, and concepts flowing daily from the world's research laboratories is liable to have far-reaching repercussions on the fabric of the world in which we live and on the texture of human relationships. In some cases this is by design; in others there are equally profound but unforeseen effects.

The social relations of science are highly complex.[1] Support for science, and the practical deployment and exploitation of its fruits, are inseparable from the counter-flow of scientific expertise into the community. Nonetheless, one can begin by considering the process from two different vantage points. In this chapter we shall look at some past and present effects of science on society, before proceeding in Chapter 6 to examine financial support for science and the motives behind it. Later in the book, the two strands will merge as we turn to contemporary questioning of the role of science in society.

First, a historical example. In 1856, at the Royal College of Chemistry in London, an eighteen-year-old chemist, William Perkin, was trying to synthesise the drug quinine from coal tar. His attempt was over-ambitious and ended in failure. In the course of his experiments, however, Perkin synthesised a purple dye – mauveine – which became the first of the entirely new series of aniline dyes. Previously, fabrics had been coloured with dyes such as indigo, extracted from the indigo plant. Large areas of Europe were given over to the cultivation of this plant. But with that single laboratory experiment in 1856, the indigo industry was doomed. Perkin's discovery thus had an enormous impact on the life, work, and economic pattern of vast areas of Europe. Perkin himself set up a small factory in Hounslow to manufacture mauveine. More importantly, he began the work of fashioning a whole range of dyes of different colours.

This was the origin of the great organic chemical industries of the world, producing a staggering range of dyestuffs, drugs and other substances. It was the dyestuff industry that spawned not only the modern drug industry, but also those making explosives, detergents, plastics, and synthetic fibres. 'This, indeed, was the birth of what we now know as technology – the application of science and the results of scientific research to the solution of practical problems: industrial, military, agricultural, medical, and organisational,' writes Lord Todd. 'It is the new technology which has revolutionised our lives in this century and which has advanced at an ever-increasing speed, fed by, and itself feeding, a similarly advancing science.'[2]

Most spectacular was the rise of the German chemical industry as a direct result of Perkin's work. In 1863, August Wilhelm Hofmann, who had been director of the Royal College of Chemistry when Perkin did his classical work there, returned to Germany, where he set about organising a German dyestuffs industry. He and his colleagues carried Perkin's work further and within twenty years the German synthetic dyestuffs and fine chemical industries had outstripped their counterparts in Britain. As Hofmann pointed out, German national characteristics were particularly suited to the development of organic chemistry, which entails a considerable amount of routine work, many slight variations of one experiment having to be performed in tracing the properties of closely related substances. Such work lends itself to highly organised, directed research.

A later example of gargantuan consequences stemming from small beginnings in chemistry research was the Haber-Bosch process. This is a method of combining nitrogen and hydrogen, by means of catalysts, to form ammonia, which is the starting point for manufacturing synthetic nitrates. As a result of the discovery of this process in 1908 by Fritz Haber, and its practical development in 1913 by Carl Bosch, production of synthetic fertilisers – and of the nitrates needed for explosives – increased enormously. Professor Kenneth Boulding, an economist at the University of Colorado, suggests that the inception of the Haber-Bosch process is the single most important event of the twentieth century. 'If it had not been for Fritz Haber,' he writes, 'the Germans would not have been able to fight that war (1914–18) because they were cut off from Chilean nitrates. Historically, there was a famous viewer-with-alarm about 1899, the English chemist Sir William Crookes, who predicted the exhaustion of Chile's nitrates and consequent global starvation by 1930. His prophecy did not pan out, thanks to the Haber process.'[3]

Many of the discoveries in science that have had the most potent practical effects on the world have been, like this one, examples of what Dr Ian Ramsey[4] has called 'moral ambiguity'. In other cases, they have been developments which, while not being capable of being turned to good or evil ends, nonetheless have created new problems as they solved old ones. The great conquests of medical science are of this sort. It is arguable how much the defeat of infectious disease over the last century was attributable to the discovery of drugs and vaccines, and how much to improved standards of hygiene and sanitation.[5] What is incontestable is that medical science has made a fantastic impact on the health of the community. In all western countries, for example, the expectation of life at birth has risen dramatically as medical science has progressed. In the United States in 1840, the figures were 38·7 years for men and 40·8 years for women. By 1960, they had almost doubled, to 75·7 and 80·1 respectively. The statistics for the effectiveness of particular techniques are just as impressive. Before 1940, when toxoid was first introduced on a wide scale in Britain to immunise babies against diphtheria, this disease afflicted 50–60,000 victims every year. After that date, the number of cases began to fall spectacularly, and since 1955 there have been only rare outbreaks. The same picture can be presented for numerous similar vaccines and drugs, and for many other countries. A speaker at a memorial service to Sir Alexander Fleming some years ago said that, through the discovery of penicillin Fleming had made a greater contribution to the relief of human suffering than any other person who had ever lived. Such achievements are immeasurable, but the dramatic efficacy of penicillin in use against bacterial pneumonia and other once-dreaded diseases needs no journalistic emphasis.

At the same time, such triumphs have undoubtedly contributed towards the greatest single problem facing the world today – that of population growth. My next two examples, therefore, are of two projects specifically designed to help in dealing with the population crisis. First, consider the contraceptive pill,[6] a unique product in the speed with which it was developed and brought into practical use by millions of women all over the world. Development of 'the Pill' began in earnest in 1956, but can be traced back to work by many different scientists studying the process of ovulation over the past century. In the early years of this century, physiologists discovered that the ovaries played a role in regulating a woman's sexual cycle, and in the 1930s they began to learn something of the hormones responsible for controlling the process. Then they found that injection of the steroid hormone progesterone would inhibit ovulation in the rabbit. However, rabbits ovulate only after mating, in contrast to Man and most other animals which ovulate at a particular point in the menstrual cycle. It seemed unlikely that progesterone would have the same effect in Man. More important still was the question of cost. Because they are complex chemicals difficult to manufacture, steroid hormones were prodigiously expensive at that time. A birth control pill for widescale use would be possible only if and when an effective contraceptive steroid could be produced inexpensively.

There are, however, steroids in plants which can be grown cheaply. It seemed possible that such steroids could be converted into others which would be effective in controlling fertility in Man. The chemist Russell Marker was the first to develop the necessary chemical technique. He used it in the early 1940s to synthesise progesterone from steroids in wild yams, which he collected personally in the Mexican jungle. In 1943, he offered his services to a company in Mexico that was later to become the Syntex Corporation, a major world concern in the fertility control business.

Marker is said to have presented as his credentials two jars filled with four and a half pounds of progesterone – worth about 160,000 dollars at the then market price. By 1950, steroid hormones were available at one hundredth their price a decade previously.

But still no-one had developed a steroid that exerted a contraceptive effect when given by mouth. Then a Syntex chemist, Carl Djerassi, synthesised a progesterone that was effective by mouth, as a possible means of preventing habitual abortion. At this point, Mrs Margaret Sanger, the American pioneer of birth control, played a key role. She had opened the first family planning clinic in the USA in 1916 in Brownsville. In 1951, she went to see Dr Gregory Pincus of the Worcester Foundation at Shrewsbury, Massachusetts, and persuaded him to slant his current research on mammalian fertility towards the possibility of an oral contraceptive. Pincus decided to pursue this line of research, and in 1954 reported that a number of synthetic steroids related to progesterone inhibited ovulation in the rat. One of them, norethisterone (known as norethindrone in the USA) also proved effective in women, and in 1955, at the Fifth International Conference on Planned Parenthood, held in Tokyo, Pincus described his first successes. The only apparent problem was occasional 'breakthrough bleeding', but Pincus and his colleagues overcame this by using a mixed pill containing other substances, which they introduced in 1957. Since then a number of modified formulae have appeared on the market, each dependent upon the original work of Pincus and his team.

Within ten years of that historic announcement at Tokyo, ten million women had taken the Pill. There was an extensive trial in Puerto Rico in 1957, and first trials began in Britain in 1960. By 1962, according to Family Planning Association clinics, 3,536 women in Britain were taking the pill. By 1963 the figure was 13,760, and the following year 44,000. By 1966, the estimates were 800,000, more than five million in the United States, and two million in the rest of the world.

No other substance has ever been consumed so widely so soon after being put on the market, aroused so much controversy, or influenced sexual attitudes so profoundly. Beginning with the work of the Roman Catholic Dr John Rock, a member of the team working on the Pill with Pincus, the Roman Catholic Church was shaken by one of its biggest rows ever, over the morality of 'artificial' contraception. But those who hoped that a non-mechanical form of contraception would provide grounds for the scholastic theologians of the Vatican to approve planned parenthood, were bitterly disappointed on 29 July, 1968, when Pope Paul's encyclical *Humanae Vitae* confirmed that there was to be no change in the position of the church. Nonetheless, the introduction of the Pill has had a momentous effect on millions of human lives, bringing release from the fear of unwanted pregnancy, particularly in those families and those parts of the world where the creation of new life is not a joyous experience but a costly burden. Coincidentally, the Pill has been blamed for its contribution to the decline of standards of sexual behaviour – just as another product of science and technology, the cheap, mass-produced car of Henry Ford, led to an increase in 'promiscuity' by making mobile seclusion available to the young in the America of the early decades of this century, and as the availability of penicillin reduced fears of venereal disease.

In the matter of medical risks, the Pill has been blamed for an astonishingly wide variety of untoward side-effects. They have all – real and imaginary – received considerable publicity. In fact, with the exception of a slight capacity to promote blood clotting in a tiny proportion of women who are already susceptible to this tendency, the contraceptive pill has proved to be one of the safest potions ever devised. It is certainly so in relation to its beneficial effects and the speed with which it was introduced. Yet it was all something of a gamble. During a BBC broadcast some years ago, Dr Gordon Wolstenholme, director of the Ciba Foundation in London, argued that 'Dr Pincus's pill was one of the most risky and foolhardy measures ever to be put into general use, particularly against an underdeveloped people such as the Puerto Ricans. The fact that it has come off is quite remarkable.'[7] He later amplified his comments as follows: 'It was a bigger risk among the Puerto Ricans, not so much because they could not appreciate the risks as because if anything went wrong the emotional racial reaction would have been overwhelmingly strong. Nevertheless, Pincus accepted the responsibility, he introduced all possible safeguards and controls, and I admire him immensely for doing it. In the history of this century . . . the introduction of the pill will turn out to be one of the most responsible things done by any one individual.'

These two comments are *not* contradictory. The successful launching of the Pill was a major scientific achievement. It was also a gamble that paid off.

The other way to tackle the population problem is, of course, by increased food production, and my second example is the introduction of new cereal strains, which, in recent years, have revolutionised agriculture in India and other developing countries. The story began in the 1950s, when the Rockefeller Foundation sent Dr Norman Borlaug to Mexico to initiate research towards new breeds of grain with increased yields. Improved irrigation and modern fertilisers had generally been disappointing when used with cereal varieties adapted to the poorly nourished soil and inadequate irrigation of the tropics. Borlaug succeeded by breeding dwarf varieties better able to use the increased water and fertiliser. As a result, Mexico's wheat production nearly doubled between 1960 and 1964. In 1962, Dr Borlaug toured the wheat-growing regions of India. In the years that followed, work by Dr Borlaug and scientists at the Indian Agricultural Research Institute in New Delhi led to the breeding and introduction of further high yield strains of wheat, and other cereals including maize and rice, able to make the most efficient use of water and fertiliser under Indian conditions. As a result, production boomed. In 1967–68, there was a spectacular harvest of 95 million tonnes, after two successive droughts and an appreciable fall in food production. The yield of wheat doubled from about seven million tonnes in 1951 to more than 15 million tonnes in 1968. In 1970, Dr Borlaug received a Nobel Peace Prize in recognition of both his scientific achievement and the potential contribution of his work to world peace.

Agricultural scientists are now building on these foundations. One aim is to develop cereals with higher protein value, rather than just increased calorie content. Because of the large quantities of cereals consumed in developing countries, even a tiny increase in the protein content of grain, or a marginal improvement in the quality of the protein, can have a considerable impact on nutritional standards and thus on health. Another goal is to boost the resistance of the new strains to disease. Some of the first high yield strains to be used proved more susceptible to infection than their natural precursors – a danger also emphasised by the understandable tendency of farmers in the early days to sow a single, high yield variety over their entire land in order to maximise their harvest. Should an infectious fungus or virus disease once gain a foothold in such an area of 'mono-culture', it could play havoc with a cereal strain that is susceptible. In an attempt to breed improved varieties, and create new mutants by exposing seeds to atomic radiation, the UN Food and Agriculture Organisation is now running an ambitious programme co-ordinating the work of different scientists throughout the world. This is another example of the moral ambiguity of science – the original research in atomic physics which has given us methods of irradiating seeds in this work (as well as atomic power stations and techniques for irradiating patients in the treatment of cancer) also led to the atomic bomb and thermonuclear weapons. It is, of course, possible for Man to decide which application(s) he prefers.

These studies also underline one of the more personal implications of the social effects of science. Introduction of improved cereal strains is highly acceptable, in a way that the foisting of new synthetic foods on a needy population is not. Scientists have devised many forms of synthetic food in attempts to help the developing countries. There are soups made out of algae grown in sewage, protein supplements manufactured by growing yeasts and bacteria in giant tanks, artificial steaks fabricated by extracting protein from leaves, and many more. But to introduce any one of these on a useful scale means gigantic changes in food preparation, and requires special channels of production and distribution. Above all, the customers have to grow accustomed to new dishes and new tastes. With high-yield cereals, on the other hand, nothing need change. The farmer sows high-yield varieties in place of the low-yield ones he grew previously. The miller and baker use the grain as before. And consumers continue to consume traditional dishes. The 'green revolution' has turned out to be by no means the unmitigated blessing it appeared in the early years, with unexpected technical and social problems,[8] but to have pioneered a totally new form of food on the same scale would have caused much greater trouble.

Communication is probably the area where science can have the most profound influence, for good or evil, on human life. Consider, for example, the immeasurable impact of television. It originated in the 1920s with the experiments of the Scottish inventor John Logie Baird, a clergyman's son who had abandoned a career in engineering because of ill-health and began messing around with his crazy idea for television after failing to make a living as a salesman. Baird devised a mechanical television system, which the British Broadcasting Corporation used for the world's first television service in 1930. But Baird's method was doomed to failure. Television is based upon a means of scanning the picture to be transmitted, so that detail is broken up into a record of different light intensities which can be reassembled at the receiving end. In Baird's system, this was accomplished by mechanical means. But electrons can scan incomparably more quickly than any device driven by an electric motor, so his apparatus was superseded with the rise of electronics. An electronic system became feasible just before the war, when Vladimir Zworykin, working at the Radio Corporation of America, introduced his electronic camera. It was then poised for development. New materials and production methods devised for other purposes during the war helped enormously, and in 1946 television spread rapidly throughout the USA. Soon afterwards, it began to grow just as explosively in other countries. When Neil Armstrong stepped on to the moon on 19 July, 1969, he was watched by an estimated 723 million people in 47 countries (over a fifth of the world's population).[9]

There can be no question of the social impact and value of television, with its enormous potential for reaching and influencing the minds of millions. On the positive side, people are better informed about their own community and about events throughout the world than they were before. They are thus more effective citizens and, being confronted by the immediacy of war, famine, and disaster on the other side of the globe, their human sympathy and sensibility can be extended and harnessed in a way that was impossible previously. Television has also affected profoundly the processes of politics and government. In the United States, for example, sustained exposure on television helped Senator Joseph McCarthy to destroy himself and his communist witchhunt in the early 1950s. At other times, glamorous exposure on the screen has had an inordinate influence on the fortunes of politicians. Television has stimulated interest in the arts, as well as being an important medium of entertainment. Its contributions to human happiness – above all, perhaps, the happiness of the elderly and the lonely in communities where they are neglected – are also very real.

Baird, Zworykin, and the other pioneers did not speculate much about any of this – they were interested above all in solving scientific problems and conquering technical obstacles. They would be staggered to know of the rate at which the medium has developed. They would also, surely, be bitterly disappointed at the shadier side of television – the rubbish, the pabulum, the profit-hunting, advertising and mental pollution that always threaten to engulf the valuable qualities of the medium.

All the more reason, then, for us to consider carefully other new communication channels that are now being created, and which – even more quickly and perhaps more extensively than television – will alter the scale and nuances of human intercourse. On the domestic front, there is the burgeoning field of videotapes. A direct descendant of television, this potent communications medium has gigantic potential. Yet its development is being dictated not by any remotely democratic process but solely by commercial manœuvres. On the world scene, we confront the deployment of satellite communication, a technical achievement stemming largely from the space research programmes of the United States and the Soviet Union. Satellite communications could help to unify the world, making it possible for people of all nations to share cultural and educational programmes. But will this happen? 'The real problems are those of attitudes, of the power drive of governments and their leaders,' writes Professor Samuel Silver. 'There is a great danger that such systems may become merely another set of instruments of international confrontations and warfare, that they will be carriers of propaganda, and media for propagating even further those aspects of our technological civilisation that already weigh heavily on our spirits.'[10]

Another, more recent, source of disquiet stemming from advances in electronics is the increasing use of computers to store and analyse personal information. Throughout the world, there are now groups dedicated to exposing the potential abuses of computers in invading privacy and severing the barriers of confidentiality that are considered important in society (see Chapter 8). Extreme protagonists would like to see the use of computers abolished completely.

The target of this criticism can be traced back to a machine invented by Blaise Pascal in 1642 for carrying out simple arithmetic. It arose out of a solid practical need. Pascal's father, a customs official, used it for doing financial calculations. There was very little progress beyond this gadget – a hand-driven affair based on wheels with ten teeth representing digits – until 1812. In that year Charles Babbage, working at Cambridge University, dreamed up an automatic machine, the 'Analytical Engine', to be used for carrying out the complex calculations necessary in compiling scientific tables. Unfortunately, the engineering methods of those days could not provide the precision required to match Babbage's blueprint. As a result, his machine – the first design for a computer in the present-day sense – was never built. Then, in 1889, Dr Herman Hollerith introduced calculating machines based on the storage of information in punched cards – the presence or absence of holes in particular positions on any one card representing 'yes' or 'no' answers to questions. Hollerith was a statistician employed by the US Bureau of the Census, and he invented his machine under pressure when it was realised that without a quicker means of sorting information, the 1880 census would not be completed in time for the next one in 1890.

The punched-card system developed rapidly after Hollerith's work. Then, in the mid 1930s, Dr Howard Aiken of Harvard University began to marry punched cards with the ideas of Babbage. By 1944, in collaboration with the International Business Machines Corporation (IBM), he had produced an Automatic Sequence Controlled Calculator – the first fully automatic calculator. As with television, crucial advances made in electronics during wartime then played their part. In 1946, engineers at the University of Pennsylvania built the first electronic computer. Like the replacement of mechanical gadgets by electronics in television, this speeded up the process of calculation many times over. Since that time, further advances have been made in the speed of computers, in their handling capacity, and in the versatility of their operations.

Hence public suspicions about 'computer domination'. The fears are of two sorts – one unimportant and resting on a misunderstanding, the other justifiable and serious. The first devolves on suspicions that computers are becoming so powerful that man can hand over large areas of life and work to electronic monsters which may get out of control. Such fears are encouraged by those scientists who talk of 'computer intelligence'. It is true, of course, that computers nowadays can be designed to control extremely complicated processes. There are, for example, factories producing antibiotics in which a computer regulates the entire interlocking series of processes. Engineers have also designed computers with active memories which simulate the processes that occur in the human brain. But the important word here is design. A computer is only as good as the information, and the instructions (the programme), it is given. However powerful the computer becomes, it remains a tool which man exploits for his own purposes.

The other fear is much more soundly based. It centres on the increasing extent to which personal information – about our medical history and financial affairs, for example – is collected and stored in the memory banks of computers used by banks, hospitals, local authorities, etc. Can such information get into the wrong hands? Is there a risk that confidentiality will be broken? What of the possibility that malicious officials will be able to correlate information between records compiled for different purposes? What safeguards are there against the possibility that the computer will make mistakes, personally damaging to the individual and perhaps unknown to him? These are all valid fears. Early in 1972, for example, the Department of Health and Social Security in Britain admitted that it had used its computerised records – previously regarded as strictly confidential – to help the police in tracing illegal immigrants into the country. And we have all had examples, usually through computerised accounting and banking procedures, of the grotesque errors perpetuated by electronic circuitry programmed by fallible human operators. Solutions to such problems are not beyond the wit of man, however. One safeguard is that any individual who provides information to be stored in a computer should receive a regular print-out (a record of the information held by the computer) to check its accuracy.

But the problem of confidentiality is not as simple as it is often represented. Take the question of medical records. There is an age-old convention that personal information disclosed to doctors and priests is given, and received, on a basis of total trust and confidentiality. Your medical history, stored away at a hospital, is supposed to be totally confidential, its contents unavailable to inquirers. Is this still true, and should it be so? The answer to both questions is no. There are many serious medical problems which cannot be solved in a hospital or research laboratory, but which may be solved by scrutinising medical records. A scientist seeking clues as to the cause of skin cancer, for example, is as likely to succeed by studying the detailed medical histories of thousands of different people with skin cancer, and seeking statistical signs of common patterns in their background, as by examining malignant skin cells under the microscope. Both approaches are important, but in the past we have tended to forget the importance of the first, partly because of spectacular successes of the second in other areas of medical science.

In recent years, however, the statistical method has come into its own – helped by the capacity of the computer to scan large numbers of records in a search for common patterns. The first proof of the link between lung cancer and smoking was possible because doctors at 25 hospitals in Britain were prepared to make available to research workers information from the records of patients with lung cancer. Likewise, our knowledge of the capacity of an early version of the Pill to cause blood clotting in a tiny fraction of women emerged from a scrutiny of records of death throughout Britain and individual records of the causes of hospital admission in an entire hospital region. There are many other examples. In contrast to thc carcfully controlled and reproducible laboratory experiment, the significant relationships in each case became apparent only when scientists studied and compared hundreds of thousands of patients' records.

Such systematic correlation of information cannot usually be strictly defined as in the interests of the individual patient. It does, nonetheless, benefit the community as a whole. And far from being the latest scientific blow against privacy, computers facilitate this beneficial work. They also make it possible to protect individual records from unauthorised or malicious access much more effectively than is possible with handwritten cards in the doctor's filing cabinet.

Science and technology in war

A fifth of the world's scientific manpower is now employed by the military at a cost of over 200 billion dollars a year,[11] and it is in wartime that a community feels the impact of scientific techniques in the most violent way. In recent years, the Vietnam war has been used as a proving ground for a horrifying array of new weaponry and destructive techniques. In 1971, for example, there was a novel development in shrapnel grenades.[12] These are dropped individually or in clusters from canisters called mother-bombs. A plane load consists of up to 2560, and each contains 200 to 500 small pellets, about five mm in diameter and weighing three to five grammes. The grenades explode and discharge their pellets at high velocity, embedding them in human flesh. Sometimes they kill outright, by tearing channels in the brain, liver and other vital organs. Often, they merely maim the victim. Two or three pellets embedded in the body can cause severe pain and paralysis. But that is not the end of the horror. At one time the pellets were made of metal. They could thus be located by X-ray, and the surgeon could remove them. Hence the new type of grenade, introduced in 1971, which produces plastic fragments. These are just as effective as metallic pellets but invisible to X-rays. The misery they cause is thus less amenable to medical attention. Scientists helped to devise this weapon. At about the same time, other scientists developed the new 'improved' Napalm B, which sticks more avidly to its blazing victims than Napalm made to the original formula. In 1972, evidence was also reported that the US forces were using scientific techniques to promote rainfall as an offensive measure.[13]

The Vietnam war was also the first conflict in human history in which one side used chemicals designed to damage and kill plants. They were provided by the United States and sprayed from the air for two purposes, to clear jungle and reduce the hazards of ambush, and to destroy food and thus starve the enemy. The chemicals involved – similar to the herbicides used in peacetime as selective weedkillers – fulfilled these roles, but also caused a variety of far-reaching, dangerous and unforeseen effects. 'Herbicides can affect the land, water, and living things which must support the people in peace or war, in independence or under foreign domination, and whatever their form of government,' wrote Arthur Galston, Professor of Biology at Yale University, in 1968.[14] 'When we intervene in the ecology of a region on a massive scale we may set in motion an irreversible chain of events which could continue to affect both the agriculture and wildlife of the area – and therefore the people – long after the war is over.'

From the very beginning of defoliation in Vietnam in 1961, scientists warned about its dangers. Not until the end of 1970, however, after a team of scientists from the American Association for the Advancement of Science had been to Vietnam and published a detailed account of their findings, did President Nixon announce the phasing out of defoliant operations. By that time, at least 17,000 square kilometres – about one tenth of the area of South Vietnam – had been sprayed with some 15 million gallons of defoliants. Over 20 per cent of the 1·2 million acres of mangrove forests in Vietnam were totally destroyed. Worse, some inexplicable factor has prevented the expected re-growth of vegetation in these areas.

The AAAS team reported that defoliation had resulted in 600,000 people being cut-off from their food supplies. They also found clinical evidence confirming suspicions that two of the principal herbicides used in Vietnam (2,4-D and 2,4,5-T) had caused birth deformations in humans, as they do in animals. Studying the records of abnormal births at Saigon's children's hospitals between 1959 and 1968, they noticed that there had been a sudden rise in the incidence of cleft palate and spina bifida after 1966, when intensive defoliation spraying began. The rate of defective births at Tay Minh Hospital, in a heavily sprayed area, was 64 per 1000, compared with 26 per 1000 in one of Saigon's better maternity hospitals. It will be many years before we know the full ecological consequences of defoliation in Vietnam.

Meanwhile, south east Asia has also been a proving ground for the most sophisticated military electronics ever devised. In the late 1960s and early 70s, the US government faced an apparently insoluble dilemma – how to withdraw forces, yet without seeming to back out or abandon their supposed responsibility to the South Vietnamese government. At the same time, the military had to sustain morale and discipline among troops who had lost heart in fighting a hopeless, unwinnable war. One horrifying yet highly effective answer to the problem came in the form of anti-personnel weapons[15] and unprecedented techniques of automated warfare.[16] As the GIs left south east Asia, electronic circuitry and advanced weaponry was deployed to detect, locate, and destroy the enemy. Computers were used to process information about Viet Cong positions and movements and calculate the most 'efficient' means of attack, which was then carried out by a battery of automatic and semi-automatic weaponry.

Developed to its highest level of sophistication, the 'automated battlefield' could leave only the major decisions of battle to the generals. For the rest, electronic detection, computers, technicians, and robot weaponry suffice.[17] Most important, computerised warfare poses no problems of loyalty and patriotism. Using electronic data-processing, obedience is unquestioned, bravery automatic. 'I personally think it has the possibility of being one of the greatest steps forward in warfare since gunpowder,' said Senator Barry Goldwater, when he first heard about the automated battlefield.

The principles behind this new dimension of warfare are childishly simple. Sensors, which can detect the presence of human beings or vehicles, are distributed over the territory, usually by being dropped from aircraft. They can be suspended in trees and vegetation or buried in the ground, and destroy themselves automatically if interfered with by the enemy. Some sensors work by monitoring vibrations caused by people walking or vehicles on the move. Others act as microphones, picking up sound directly. One of the most advanced, the 'people sniffer', detects the vanishingly small quantities of ammonia in human perspiration. An automatic command centre receives information transmitted from a network of such devices. The computer processes the information and gives the results to an officer or 'skilled target analyst'. It becomes the basis for activating a similar battery of weapons – including rockets, buried mines, aircraft carrying laser-guided bombs and a range of other bizarre weapons.

One of the first practical applications of the system was against the Ho Chi Minh trail. A pilotless aircraft, the QU 22B, was used to forward data from field sensors to the Nakhom Phanom base in Thailand – reportedly one of the largest telecommunications centres in the world. There IBM computers processed the information, and instructions for offensive action were passed to field commanders, bomber bases, and 'telecommanded' mines. Satellites controlled from the Nakhom Phanom centre were also responsible for radio listening and other observations over China and south-east Asia. Another satellite linked the base directly to the Pentagon.

In a typical operation of the automated battlefield, the duty officer of a particular sector sits at his control panel watching a screen which displays data from the network of sensors in his area. Like a radar screen, it gives a minute-by-minute picture of the territory, which may at that moment be shrouded in darkness. Noting evidence of activity in one area, the officer can order an attack. Alternatively, he may arrange for his computer to send information to another computer on board an aircraft. The second computer then calculates the course and altitude for the plane to release its bombs at precisely the right place and time. A third possibility is that, by a series of judicious but moderate attacks, the officer can use his gadgetry to herd the enemy into one particular area for complete elimination. The system was used in 'free fire zones' – areas where, following the dropping of warning leaflets, it was assumed that all people were enemy combatants. But because people often refused to leave their homes, this assumption was invalid. Moreover, the sensors cannot distinguish between soldier and peasant, between man, woman, or child. For this reason, and because the officers of the automated battlefield can work without ever seeing the enemy, the system is totally indiscriminate.

Unlike the 'automated battlefield' but like defoliants, which were introduced originally for the peaceable business of weed-killing, the harassing agent CS was the outcome of non-military research. It was developed further through a research programme in the early 1950s directed towards improving the substances then used to control riots. According to a paper written at the time by scientists at the UK Chemical Defence Experimental Establishment, at Porton Down, the most widely used agent, CN, though it irritated the eyes intensely and had other unpleasant effects, was 'not adequate for incapacitating or even seriously discouraging fanatical or highly motivated rioters'. The Porton scientists studied many possible replacements, and turned eventually to a substance,

CS, which two chemists, Dr B. B. Corson and Dr R. W. Stoughton, had described in the *Journal of the American Chemical Society* in 1928. They had synthesised CS during the course of fundamental research into chemical reactions. They noticed that it behaved like a tear gas but were not themselves interested in pursuing its practical application further.

The Porton group transformed CS from a laboratory curiosity into a practical harassing agent. Following its introduction in 1962, police and the military used CS to quell demonstrations and riots in many parts of the world – from student demonstrations in Berkeley and Paris to civil disturbances in Ulster. At the same time, disquiet and opposition to its use was growing among scientists and doctors concerned about its ill-effects.[18] In Britain, this culminated in the setting up of a committee under Sir Harold Himsworth to consider the use of CS in Ulster. Its final report,[19] which appeared in September 1971, recommended some minor changes in the way CS was deployed, but advised the Government that its continued use was safe.

It is, of course, easy to condemn totally the use of CS gas and anything like it, at all times and in all circumstances, as an abuse of civil liberties, an affront on the individual. Nonetheless, there are occasions when crowds behave in an ugly, violent way, and when canisters of a harmless incapacitating agent are more sensible weapons to deploy than firearms. The argument should therefore be over whether or not CS is the harmless agent it is supposed to be. By definition as a harassing agent, CS is a vicious chemical with extremely unpleasant effects on the body. But is it safe to use on crowds of rioters and innocent bystanders; on the mixture of old and young, fit and ailing, who will inevitably be exposed; in all possible weather conditions and physical environments; on repeated occasions on the same victims?

These are the questions that should be answered fully, and in the affirmative, before an agent such as CS is made available to the police or military for use in controlling disturbances. They are much more stringent safeguards than would be required for a chemical agent unleashed in more extreme but more restricted circumstances. In 1968, the Under Secretary of State for Scotland said in a Parliamentary answer that in Britain such weapons as CS were to be used only against 'armed criminals or violently insane persons in buildings from which they cannot be dislodged without danger of loss of life, or as a means of self defence in a desperate situation, and that in no circumstances should they be used to assist in the control of disturbances'. The intensive application of CS in August the following year in Londonderry was in flat contradiction of this statement – a profoundly disturbing situation for those concerned about the channels through which a community decides how to deploy the fruits of science.

The Himsworth report was far from satisfactory. It was not reassuring to learn that 'it is only under quite exceptional circumstances' that one could receive an excessive dose of CS, causing death. Similarly, the risk of serious eye injury caused by CS pellets lodging in the eyes was described as 'small', and diarrhoea in babies following exposure was said to be 'of no practical importance'. The report did, however, confirm that inhalation of the gas could provoke acute exacerbation of bronchitis and asthma – and possibly of heart disease too, though the evidence on this last point was 'inadequate'. Finally, 'a more definite assessment of the risks' that CS could cause cancer had to await animal experiments still in progress. Meanwhile, the committee had some practical tips. When a pellet grenade was used, for example, 'there should be a warning to rioters to keep their eyes closed'.

Particularly disquieting is the fact that CS was used in Northern Ireland in far greater quantities than envisaged by either the laboratory scientists who developed it as a riot control agent or the authorities who had sanctioned its use. It was deployed in an indiscriminate, punitive manner, and in circumstances where elderly or ailing bystanders could not escape (as Himsworth suggested) or even find sanctuary in their own homes. The experimental evidence about the effects of CS on the human body came from work with service volunteers. They were fit, healthy young men, aged 18–30, who received £1 per minute for being exposed to a strictly controlled concentration of CS under defined laboratory conditions. That was not the situation in Northern Ireland. Moreover, of the 48 references used to support the report, 22 were to secret Porton papers or 'personal communications'.

Science in sport

No sector of life is unaffected by science. It even plays a number of roles in sport, one of which affects the very basis of sporting competition. The 1968 Olympic Games, held in Mexico City, marked a crucial turning point in the application of science to sport. There, and again in Munich in 1972, doctors and physiologists were busy scrutinising competitors, counting the chromosomes of lady athletes, and collecting urine samples at the finishing line in an attempt to detect 'drug taking'. The purpose behind all this activity was the elimination of unfairness in what should be open competition between the natural skills of competitors. Ostensibly female competitors whose chromosomes in fact more closely resemble those of a male should not be allowed to compete in ladies' events, and no competitor should be allowed to consume drugs to boost performance. Yet officials apparently turn a blind eye to some practices, such as the use of anabolic steroids to promote muscle bulk in weight lifters. Just before the 1972 Olympics, Dr Pat O'Shea, exercise physiologist and a member of the US Olympic Weight Lifting Committee was quoted as saying: 'If we were informed we could not select an athlete taking steroids, we simply wouldn't have a team.'[20]

The possibility of detecting sporting 'offences' is a direct result of scientific advances in two different fields. Comparison of human sex chromosomes became possible only in the late 1950s when cytogeneticists learned how to prepare and stain human cells in such a way as to visualise the chromosomes clearly under the microscope. This was a considerable technical achievement, as for many years previously methods were so poor that biologists believed that there were 48 chromosomes in human cells. Then, in 1956, Joe-Hin Tjio and Albert Levan, working at the University of Lund in Sweden, developed improved techniques and discovered that Man had only 46 chromosomes. The detection of drugs in athletes' urine was made feasible by the technique of gas chromatography, developed originally in 1953 by A. J. P. Martin and A. T. James, two Medical Research Council scientists in London, as a means of analysing vanishingly small amounts of organic substances. In both cases, techniques continue to improve. Chromosome analysis at the moment can reveal only major abnormalities and differences between individuals – the presence of an extra chromosome, for example. But in the future methods will be available of discriminating very much finer shades of difference. Likewise, the range and sensitivity of gas chromatography is always increasing.

The application of science to the problem of fairness in sport seems doomed to failure. Thanks to advances in many different branches of science – particularly physiology and nutrition – we now know incomparably more about getting the best out of the human body than did the Greeks who set up the Olympic ideal. Coaching has become highly scientific. Football trainers use all manner of gadgetry to get their players fit and keep them so. Diets are carefully selected for the man and the event. The same freedom to boost an athlete's talents should apply to drug taking – subject, of course, to the fact that athletes, and anyone else for that matter, should be discouraged from consuming *dangerous* substances. It is, in any case, impossible to impose a logical barrier between what is and what is not a drug. Before the Munich and Mexico City Olympics, competitors were handed a long list of forbidden substances which they must not use. It included narcotics and stimulants of the nervous system, such as cocaine, but also more innocuous substances including anti-depressants and alcohol. The marathon runner was allowed to relieve a nagging headache with aspirin, but should he suffer a bout of depression, caused by worry over his performance, he must not take the relevant medication. The sprinter who fancied a double Scotch to steel himself for the great occasion must not imbibe, though he could (with more effect!) pep himself up with caffeine from several cups of strong black coffee. These are absurd anomalies. But unless sports authorities decide that 'anything goes', even more scientific expertise will be devoted to, and wasted upon, sport than at present. Working against the scientists who are so anxious to detect outlawed drugs will be pharmacologists trying to devise new drugs that boost performance yet remain undetectable by the techniques used by officialdom. An absurd prospect.

Two myths debunked

I have not tried, in this chapter, to present a comprehensive account of the impact of particular sciences on society. Instead, I have chosen a variety of examples of tangible, far-reaching, and in some cases unexpected effects of science on different areas of society and human life. They also illustrate the massive *scale* on which new discoveries can be deployed, marketed, and in other ways made to impinge on the lives of millions of people soon after leaving the laboratory or drawing board. In many areas, the scale on which this happens is enlarging all the time. It is important to distinguish this from the widely accepted myth that the rate of technological innovation has increased in recent years. Several studies have purported to show such an acceleration in the pace at which discoveries progress to practical adoption, but the case is far from proven. For every analysis apparently showing a decreasing time lag between discovery and practical application, another can be constructed proving the opposite.[21] While much depends on the selection of examples, there is also an inbuilt bias which tends to accentuate the rate of change in recent years. Even with no trend whatever in the rate of adoption of inventions, a study which looks only at average lag times for different periods is almost bound to find an increase in the rate of adoption for the more recent periods, because of the omission of recent inventions which have not yet reached practical fruition.[22] Perhaps the greatest source of error, however, is the selection of starting points for particular innovations. One widely published chart, used many times in magazine articles and books, is particularly suspect in this regard. It shows photography spanning 112 years from discovery to practical adoption, and the more recent invention of the atomic bomb taking just six years. But the starting point chosen for photography is 1725, when someone first noticed that silver salts darkened in response to light, while that for atomic weapons is the beginning of the Manhattan Project. Clearly, the two events are not comparable. With equal logic, one could date the beginning of photography much later, with the first permanent photograph (1822); that of atomic weapons could be placed much earlier, towards the beginning of the century, with early experiments and speculation over the structure of the atom and the interchangeability of matter and energy.

Finally, there is another myth to be discounted. I have not in this chapter always distinguished between the laboratory science behind particular discoveries and the development work that carried them to practical fruition – between research and development, 'pure' science and 'applied' science, science and technology. I have ignored these fashionable distinctions because they suggest a far too rigid separation between the two phases (whatever the labels). Even as late as the last decade, we all believed in a simple, rigid, linear model for scientific innovation. Fundamental research led to applied research, which led to development work and eventually to new products and processes, which in turn stimulated the economic growth of industries, companies, and nations. Largely due to intensive study by research workers in the Department of Liberal Studies, Manchester University, this concept has now been shattered. The model breaks down completely in many cases. Technology, geared to social or commercial need, stimulates basic science almost as frequently as basic research acts as a seed-bed for technology.[23] The interaction between science and technology is, in fact, complex, and not always linear. Even the traditional distinction between pure and applied science is scarcely worth maintaining. Research which for one person is a 'pure', disinterested investigation can also be 'applied' research from another person's standpoint.

Far more useful is the contrast between 'curiosity-oriented' and 'mission-oriented' research – self-explanatory terms which relate research clearly to the motives of its practitioners. And this brings us to a crucial dilemma behind the financial support for science. How much money should be allocated to the fulfilment of curiosity and how much in the pursuit of hard returns?

Chapter 6
The paymasters

'However distinguished, intelligent and practical scientists may be, they cannot be so well qualified to decide what the needs of the nation are, and their priorities, as those responsible for ensuring that those needs are met.'

Stirring words. They come from a report, entitled *A Framework for Government Research and Development*,[1] which landed like a bombshell in the midst of Britain's scientific community in November 1971. They formed the text for the author of the report, Lord Rothschild, to make an outright attack on the financing of certain types of scientific research in Britain. Within hours of publication, there was uproar. Without stopping to read the report fully, scientists blasted off letters to the press, and uttered unaccustomedly strong words to newspaper reporters and radio and television interviewers. They felt threatened. In particular, they believed that their sacred freedom to follow their noses, pursuing research for its own sake, was in jeopardy. In the ensuing weeks, more than fifty letters from working scientists attacking the Rothschild report appeared in *The Times*. It was probably the most concentrated outpouring ever in Britain on the subject of scientific policy. Lone support for Rothschild came from the Confederation of British Industry and a mere handful of working scientists.

Before we consider in more detail the Rothschild report, and the reaction of the scientific community to it, we should examine briefly its context and historical background. The Rothschild debate symbolised a dilemma which has begun to face Britain, the United States, and other advanced countries in comparatively recent times. In short, why do we support science? What are our motives for doing so? Do we expect quick returns or longer term benefits? Should we, indeed, expect pay-off at all? Or is science a special activity, like the arts, which any civilised community ought to support, regardless of cost or practical outcome? In the industrialised West, we have tended to ignore such questions until very recently and were content to go on increasing scientific expenditure each year, without scrutinising our motives for doing so. In the United States, for example, total expenditure on research and development rose from less than 600 million dollars in 1940 to 22·6 billion dollars in 1966 – and from 0·6 per cent of Gross National Product to over 3·0 per cent. But during the 1960s, largely as a result of financial stringency, spending on science came under attack. Politicians, with their vague belief that science should stimulate economic growth, became disillusioned when they looked for hard evidence of a cash pay-off. The community at large became concerned too. People were anxious to see more science applied to immediate practical and social problems, and at the same time they grew angry about the ill-effects of science in the form of pollution and environmental rape (see Chapter 8). Disappointingly, many scientists reacted with a mixture of obstinacy and arrogance to this questioning. They took the line: Science is good for you. But unless you are a scientist you can't possibly understand why. Therefore scientists must be left alone to get on with their work. Q.E.D.

Hence to the Rothschild bombshell. Lord Rothschild was concerned with just one area of research spending – applied research and development. This he defined as R and D with a practical application as its objective – the development of a new antibiotic, for example, or a process for purifying uranium. Such work, Rothschild argued, should be done on a customer-contractor basis. 'The customer says what he wants, the contractor does it (if he can), and the customer pays.' In practical terms, this threatened some of the money spent by Britain's research councils – at that time a total of about £109 million per annum. Instead of being handed out to the research councils with no strings by the Department of Education and Science, Rothschild argued, money for applied R and D should be channelled through the relevant government departments. Whitehall scientists – scientific administrators – should define practical problems requiring solution, and hand over the projects (and the money) to research council scientists or any other alternative research organisation that seemed better fitted to pursue the work in question. In hard cash,

£5·6 millions was to be slashed from the annual Medical Research Council budget of £22·4 million and transferred to the Department of Health and Social Security. The DHSS would then pass on the money in payment for contract research – such contracts being placed in the first instance, but not necessarily for evermore, with the MRC. The same system would apply to other research councils and their corresponding government departments, and 'the research councils should not have the right to reject such contracts without good reason agreed with the sponsoring executive department'. It would be up to the government departments to ensure that they got what they wanted from the research scientists. 'If they do not, they can, and doubtless will, go elsewhere with their money, to get their objectives met,' Rothschild warned sternly.

Once the squeals and tremors of rage from the scientific community had abated, it became possible to discern the solid arguments on both sides for and against Rothschild's proposal. At the centre was a sizeable argument about words, resembling the more opaque inanities of scholastic theology. It hinged on the old-fashioned distinction between pure and applied research. Rothschild's intention was to leave severely alone the whole field of 'pure' research, but to change the system for financing applied science so as to obtain better value for the customer. Unfortunately for administrative tidiness, there is no rigid distinction between these two sorts of science – though there are clearly separate ends to the spectrum – and it thus became easy for opponents to create a smokescreen of words to imply that Rothschild-type accountability would be damaging. Pure science is research pursued for its own sake. It is curiosity-orientated work done in an attempt to learn more about the behaviour of matter or the structure of the universe, for example. As with the example of Dr B in Chapter 1 (see p. 21), this type of science is propelled principally by its own internal logic and the skills of the scientist himself. There is no long-term practical objective. At the other pole are projects pursued towards defined practical, political, or commercial goals. They comprise what American scientists call 'mission-orientated', as opposed to the former brand of 'discipline-orientated', research. Our microbiologist, Dr A (p. 21), was engaged in mission-orientated research, with the hope of his company marketing a profitable anti-viral drug. Here research merges imperceptibly into the development work that would take the discovery from Dr A's laboratory bench, through various stages of testing and product improvement, to a saleable commodity. Hence the 'development' of Rothschild's 'applied R & D'.

Medical research is one area where there can be no rigid compartmentalisation into pure and applied aspects. Developments of practical importance in cancer treatment and diagnosis, for example, have come from pure research with no objective other than to increase knowledge of how cells behave – 'pure' biology. On the other hand practical developments, and pressures from practical needs, can also necessitate what would otherwise be indistinguishable from pure research programmes designed to increase fundamental knowledge. Between the two extremes, there is a continuous interplay.

One fear of Rothschild's opponents was that a customer-contractor relationship would be unworkable in many areas of science having practical implications – in medicine and agriculture, for example. So it would, in some particular fields. It would be impracticable to put 'a cure for cancer' out to tender, because it would be impossible to specify the requisite work with any exactitude. We do not know enough about cancer to be able to define – and cost– the research programme that would be necessary to solve the problem. However, with many other projects there is a vast body of experience from tackling similar problems in the past to guide future plans. If a government department were to sanction the development of an oral vaccine against chickenpox, for example, it would be entirely feasible to assess the practical and theoretical problems that would have to be solved, by comparing the likely obstacles with those encountered in developing vaccines against other infections in the past. The bigger the project, the greater the inherent uncertainties, but such work could undoubtedly be specified and costed realistically.

Another objection to Rothschild, brandished freely throughout the British national and scientific press in the winter of 1971–72, was the necessity for scientists to enjoy the freedom that is vital if scientific creativity is to thrive. After initial reactions to the report, when spokesmen arrogantly assumed that the great achievements of Britain's research councils were self-evident and did not require discussion, research council protagonists turned to writing articles listing these achievements and hinting that such work would atrophy if Lord Rothschild had his way. In a paper in *Nature*[2] in January 1972, one distinguished scientist cited the discovery of the antibiotic cephalosporin under the auspices of the Medical Research Council in 1953 as a great advance which had not only saved lives but also earned £9 million in sales. But such evidence proves nothing about the necessity of freedom or research council autonomy. The discovery of cephalosporin can be matched by the development of the synthetic penicillins by Beecham (see p. 111) a few years later. This was work of at least equal practical significance, but pursued under the most urgent pressure from commercial motives. In any case, with a total scientific staff of some 5700, it would be odd indeed if the research councils could not compile an imposing catalogue of achievement. The real test of the administrative context of research is whether scientists can repeat great achievements a second time. As it happens, neither the MRC nor Beecham have yet done so in this particular field.

Many correspondents in *The Times* highlighted pure research as the seed-bed from which applied work flows. They then suggested that this would be threatened by Rothschild's proposals. Professor Sir Charles Dodds, for example, wrote of his work in the late 1920s and mid 1930s on the hormones regulating fertility in mammals.[3] Although starting as a 'pure' investigation into the relationship between the shape of hormone molecules and their activity in the body, this research yielded important practical results. In 1938, Dodds, Sir Robert Robinson, and their colleagues produced the substance diethylstilboestrol, which has since been used extensively to help menopausal women and to treat prostate cancer in the male. It has also been employed extensively in animal nutrition. 'No one could predict the production of stilboestrol,' wrote Sir Charles Dodds. 'In my opinion, it would be a retrograde step to make such type of investigations impossible.'

In fact, this type of basic, speculative research was not the target for Rothschild's report. It is as well that we remind ourselves of the potential practical value of such work, even though, as we saw in the previous chapter, the link is not as important as it once seemed. But that is a separate issue from the question of securing better value from defined projects in applied research and development. More important is the argument about what proportion of money can realistically be allocated à-la-Rothschild, on a customer-contractor basis. Experience in the United States suggests that some research even in such tricky and uncertain fields as cancer research *can* be promoted in this way – in marked contrast to the view of most British scientists, who feel instinctively that such arrangements cannot work. However, scientists and administrators on both sides of the Atlantic are still learning – indeed, in many ways just beginning to learn – about how best to finance research and exploit its results. That being so, it was as well that, when the British government took action on the Rothschild report in July, 1972, the protagonists on both sides found themselves faced with a compromise plan. A greater proportion of money than in the past was to flow into research via government departments, though less than the large slice proposed originally by Rothschild.

Before leaving the Rothschild debate, there are two related dangers we should consider regarding the direct sponsorship of scientific research by a government department. They concern secrecy and political independence. Mrs Shirley Williams, Minister of State at the Department of Education and Science in the Labour government from 1967–69, highlighted one problem in her 'party is ended' article in *The Times*[4] in February, 1971 (p. 13). Supposing, she argued, that research were to reveal dangers in the use of certain pesticides, which nonetheless were financially beneficial to farmers. In such a case, it would be politically simpler if the Minister under whose auspices the research had been conducted were not subject to a conflict of interest between safety and finance. Such conflicts could lead to the suppression of research data or to unhealthy personal conflicts within a government department. Far better that the various protagonists argue their points in the open. This was what happened in Britain in the late 1960s, when a government committee was set up under Professor (now Sir) Michael Swann to investigate public health dangers arising from the use of antibiotics in agriculture (see p. 137). It is questionable whether the Swann committee would ever have been convened if the relevant research had been conducted under the Ministry of Agriculture, Fisheries, and Food, with its primary brief to feed the nation as economically as possible. The task of a scientist under the Min. of Ag. is to solve problems for his masters, not create them.

As if to underline this type of danger, just before the Rothschild debate began in the winter of 1971, Dr Kenneth Mellanby, director of the Monks Wood Experimental Station, run by the Nature Conservancy, warned scientists publicly that government departments were censoring scientific results when these were embarrassing. Speaking at a meeting of the British Association for the Advancement of Science, held in Swansea, he described how a paper submitted to the journal *Environmental Pollution*, which he edited, had apparently been censored to save inconvenience to a particular government department. Neither commercial nor military security were threatened. The tendency to impose such censorship was increasing, Dr Mellanby felt, and it could spread to the research councils. 'A scientist in a research council must never be so muzzled. Since the research councils were put under the Department of Education and Science, there has been a tendency for scientists to be told that they must behave like civil servants and not say or do anything which might embarrass the Minister.'

A disquieting example of censorship originating in a government department, but spreading outwards from it, was given by Dr Tom Margerison, then science editor of *The Sunday Times*, in November 1966. While reading advanced proofs of *Nature* one week some years earlier, he had noticed a short article on Turkey X disease, a condition affecting young ducks and turkeys. Research at the Tropical Products Institute in London had established that the disease was caused by a toxin produced by a fungus which grew on groundnuts fed to the animals. Further tests showed that the same toxin produced liver cancer in rats. Knowing that human liver cancer was particularly common in Africa, Margerison decided to write a report on the work for his paper. 'The next day, when *Nature* was published, the article was not there,' he writes. 'I spoke to the editor on the telephone. He told me that he had been forced to withdraw it, at considerable inconvenience. I followed this up and spoke to the Director of the Tropical Products Institute. He would say nothing, but before long I found myself in a high-powered meeting at the Department of Scientific and Industrial Research.[5] Foolishly, I allowed myself to be told the whole story in confidence, and thereby I was effectively muzzled. The enormity of the situation only struck me later.'[6]

In fact, at that time government scientists suspected that the fungus responsible for turkey X disease, growing on groundnuts in Nigeria, could also be spreading liver cancer among humans who consumed margarine made from the nuts. It emerged only later that the manufacturing process prevented the toxin from passing into margarine. Nonetheless, 'the reason for the high-level conference and the hasty discussions was commercial and political. The British government did not want to make public the results of scientific research which could upset their relationship with groundnut-growing countries in Africa.'

Why the paymasters pay

Let us turn now to the positive motives of the paymasters. Why do societies, governments, companies, support science? The reasons are not always as obvious as may appear. But war-mongering is very high on the list.[7] There is, for example, tremendous research activity now centred on the oceans of the world. Among the reasons for this are hopes of exploiting 'the last resource'[8] as a reservoir of food and a source of valuable minerals and metals. But the largest amount of spending on oceanographic research is by the military, and lurking behind many an innocent-sounding research programme is the military interest. 'Twas ever thus. Galileo was Professor of Military Science at the University of Pavia when he sold his invention of the telescope to the Venetian Signory 'solely on account of its value in naval warfare'. Yet even this was not a straightforward sale.[9] The telescope was given to the Doge as a free gift in return for 1000 ducats and a professorship for life.

As J. D. Bernal points out,[10] the great technical developments of the eighteenth and nineteenth centuries, in particular the large-scale smelting of iron by using coal, and the introduction of the steam-engine, were direct results of the needs for artillery, demanded by the increasingly large scale of war. 'The accurate boring of steam-engine cylinders, which made all the difference in practice between the efficient engines of Watt and the earlier atmospheric engines, was due to the improvements introduced by

Wilkinson, who was able to make them on account of his experience in the boring of cannon. From the same field comes Rumford's discovery of the mechanical equivalent of heat, which was to furnish the basic theory for all heat engines.'

The political history of the world since the last war has been intimately associated with a grotesque and unprecedented level of spending on military science and technology. The most recent twist in the dizzy spiral began in 1957, with the launching of *Sputnik 1*, the world's first man-made satellite. The shattering implication for the United States was that the USSR had constructed a rocket sufficiently powerful and accurate to serve as an Intercontinental Ballistic Missile. At that time *Atlas* – the first American ICBM – was not due 'in service' for another two years. Hence the term missile-gap. It was invented by hard-nosed American technologists, who held out to President Eisenhower on the one hand the threat of nuclear annihilation for America and on the other the promise of increasingly esoteric, wondrous technical developments to wrench back the lead in defence technology. By 1962, a combination of reconnaissance satellites and *Minuteman* ICBMs had given the United States a commanding advantage in strategic deterrence. (As Neville Brown points out,[11] this does much to explain both the origin and the outcome of the Cuban missile crisis of that autumn.) The United States civil space budget rose from 500 million dollars in 1956 to four billion dollars in 1961, and by early 1961 President Kennedy was able to commit his country to putting a man on the Moon 'in this decade'.

Dr Herbert York, who served both Eisenhower and Kennedy as an adviser in the Defense Department, has argued[12] that developments in military technology in recent years have proved the wisdom of Eisenhower's warning that 'public policy could itself become the captive of a scientifico-technological élite'. Weapons systems have become so complicated, following successive developments in the Byzantine technology urged by the hard-sell technologists, that we shall soon have 'a situation in which the response to a hypothetical future attack will be so complicated, and the time in which to decide what to do will be so short, that it will be necessary to turn to automatic computing machines for the purpose'. The decision to initiate the nuclear annihilation of life on the planet is passing from 'the hands of statesmen to lower-level officers and ultimately computing machines and the technicians who program them'.

According to Dr Frank Barnaby, director of the Stockholm International Peace Research Institute (SIPRI), the security of the two superpowers has been decreasing steadily since the early 1960s, when there was a relatively stable strategic balance between East and West.[13] 'The ultimate folly,' Barnaby writes, 'will be attained by the extensive deployment of multiple independently targetable re-entry vehicles (MIRVs), and sophisticated missile-detection systems.' Another development leading us towards the Kafkaesque danger of computer-initiated annihilation is that of anti-ballistic missile systems (ABMs), such as the US Army's *Safeguard* system, which is now being deployed at an estimated cost of 13·4 billion dollars by 1974.

Against that sort of money, all other spending on scientific research and development pales into insignificance. It was this grim fact, coupled with a starry-eyed faith in 'the scientific revolution' as a means of changing society and creating national wealth, which led Harold Wilson, then leader of the British opposition, to espouse science so dramatically in a speech at the Labour Party's annual conference at Scarborough in 1963. 'The talents of scientists and the skills of tens of thousands of technicians in the great government research institutions have been largely squandered in the vain pursuit of Nuclear Greatness,' proclaimed the party's policy statement. 'The application of science and technology to peaceful industry and to public services such as health, transport and education, has been neglected.' The future of the scientific revolution had been left to an unplanned system of private enterprise; conservatism (with a small c) had delayed the introduction of new technology into industry; and too few industrial leaders understood science and technology. This was the main reason why British industry had 'so lamentably failed to maintain Britain's position in world markets'. The solution, urged by Mr Wilson in a memorable speech, was to forge a new Britain in the 'white-heat' of scientific revolution.

Immediately after winning the 1964 election, the new Labour government altered the organisation of science and technology in Britain,[14] creating the Ministry of Technology, and a Council for Scientific Policy to advise the Secretary of State on the budgets for a series of new research councils dealing with science, social science, and environment, and the existing Medical Research Council and Agricultural Research Council. Michael Stewart, a classics schoolmaster, became Secretary of State for Education and Science; Frank Cousins, the trade union leader, was the first Minister of Technology; and Roy Jenkins, historian and economist, became Minister of Aviation. In 1966, Sir Solly (now Lord) Zuckerman, who had been chief scientist in the Ministry of Defence, became Chief Scientific Adviser to the Government.

There was no revolution. Because of a lack of clear policy directives, even the distinguished scientists brought into Whitehall in various capacities – Lord Snow and Lord Bowden as junior ministers, and Patrick Blackett as chief adviser to the Ministry of Technology (Mintech) – soon became discouraged and returned to their former posts. Perhaps the only lasting achievement from this period was the creation of the Mintech, which strove not to develop a socialist approach to technology, but to make free enterprise based on science work profitably and aggressively. In its early years, Mintech brought about an amalgamation of the British computer industry, with the aim of helping it to stand up to competition with IBM and the other giants of the world computer business. Under the influence of Anthony Wedgwood Benn, the Minister from 1966 to 1970, Mintech began to encourage research and development projects likely to be profitable and to contribute to national economic growth, rather than 'prestige' projects. With Benn at the helm, this policy was pursued with great panache and determination. It must, the pundits argue, have had positive effects. Sadly, we really cannot be sure.

What we do now know is that the relationship between research and development expenditure and national economic prosperity is by no means as clear as it seemed when Harold Wilson made his Scarborough speech in 1963. In the intervening years, there have been innumerable reports and analyses of R and D spending in different countries, and politicians have flourished figures to attack one country for spending too little and praise another for its great wisdom. Statistics included in the Rothschild report, for example, show the United States heading an international league table of public money spent on R and D per annum. Belgium is at the bottom of the list and the United Kingdom about the middle.

What we should deduce from such figures is not clear. The countries differ in many ways. They have widely differing populations and social needs, and are at different stages of industrial development. In any case, Rothschild argued, we can draw no inferences from current figures because it takes an estimated seven years at least before R and D effort has a tangible effect on a country's welfare. Above all, we simply do not know whether an increase in GNP is in part caused by an increase in R and D expenditure. On the contrary, it may be that the latter is a consequence of greater prosperity. We know very, very little about the contribution made by science, if any, to economic development.[15]

The same is not true of the single, massive, prestige project, which produces tangible effects for all the world to see. The *Apollo* moon programme is the outstanding example of recent years. It was pursued for reasons of politics, glamour, and prestige, thinly disguised as science. Just how threadbare was the scientific justification for the manned moon programme is clear from an article by Captain Alan Bean (who himself landed on the moon in November, 1969) which appeared in the UNESCO journal *Impact*[16] just before the 'lunar rover' Apollo mission of July, 1971. Although the title of the article was 'The value of manned flights to the moon', Captain Bean quickly sidestepped the obligations of that title and talked mostly about space research generally. Satellites had been sent up, he said, to facilitate communication between nations, to provide information about weather conditions, and to survey the earth's resources. Man's horizons had been widened, 'adding new dimensions to our economic, cultural, and spiritual potential'.

None of this, of course, has the remotest thing to do with putting men on the moon. The undoubted and worthwhile benefits of space exploration, particularly for communications, do not depend in any way on the quite separate exercise of sending men to the moon. The quest to put man on the moon has actually distorted the scientific motives behind the space programme, forcing massive funds to be used in fabricating the life-support systems needed to sustain human life in the hostile lunar environment. Bean mentions the moment when, on the *Apollo 11* mission, Neil Armstrong 'took over the controls' to guide the module down smoothly on to the moon's surface. He ignores Russia's demonstration that men are *not* necessary for this type of manœuvre. Soviet space chiefs have been content to use their expertise in designing reliable automatic landing and sampling equipment, avoiding the massive extra cost of organising televised high-jinks on the moon or Sinatra sessions in space. Moreover, the type of scientific information gleaned from the moon – the haul has been embarrassingly puny in relation to its cost – has been obtainable by either human observers or robot gadgetry.

To their credit, many top-level scientists resigned from the *Apollo* programme when they realised just how low science rated in the priorities behind the missions. They included the chief scientist, the director of the Lunar Laboratory, the principal investigator of *Apollo* lunar surface geology, the curator of lunar samples, and two of the scientists who were training as astronauts. Dr Eugene Shoemaker resigned in 1970 'out of deep concern for the direction of the nation's space goal', and described *Apollo* as 'a poor system for exploring the moon. . . . The same job could have been done with unmanned systems at one-fifth the cost three or four years ago.'[17]

Captain Bean also paraded the 'spin-off' argument, embarrassing in its nakedness, explaining carefully that conditions of weightlessness might help man to manufacture more perfect ball bearings and other vitals. Apart from non-stick frying pans, he pointed out, the space effort had already spun-off such things as telemetry units, designed to monitor bodily functions in astronauts but also useful in the intensive care wards of hospitals, and a new type of hammer developed for building the mighty *Saturn V* rocket. Again, none of these developments, real or dreamed-of, have any necessary connection with moon-going. It's possible that the unique conditions of near space may prove useful in various commercial manufacturing processes, but if so the discovery will not be a result of the moon race. And the suggestion that the *Apollo* programme has in any way justified itself by stimulating science and technology is simply fatuous. *Any* large scale technological venture, however absurd, is bound to produce something useful. If the object is to make telemetry sensors and superior frying pans, one can set out to do just that – and save a lot of money.

Sadly, at the time Bean published his article, there was a bitter reality behind all that proud talk of the stimulus to advanced technology. Throughout the United States, specialised research and development laboratories, spawned during the halcyon days of *Apollo*, lay idle. The US space programme had begun to move into a more sensible phase, away from adolescent passion for the theatrical, and many of those scientists and engineers who provided the skills for the earlier *Apollo* period were no longer needed. Starting in the early 1960s, some 200 American universities had signed up with the NASA Sustaining University Programme which provided financial support for research projects. But the bubble burst all too soon, and with savage ferocity. From 45 million dollars in fiscal 1965, the programme declined to 30 million dollars the following year. The Johnson administration, facing the gargantuan costs of Vietnam, requested only 10 millions for 1967. The programme carried on at that rate for the next few years. Then, in 1970, Nixon's budget called for its termination, leaving the American academic landscape covered with decadent symbols of the salad days of lunar enthusiasm. 'For a good many universities,' wrote Dan Greenberg at that time,[18] 'the results are to be found in the form of staffs that were hired in the expectation that the NASA bankroll was inexhaustible; expensive buildings; and multitudes of space-oriented Ph.D.s for whom job prospects are about as plentiful as *Apollo* launchings these days.'

The story of Concorde is an equally sorry tale.[19] Concorde was not planned because anyone seriously wanted it – least of all the world's airline chiefs who, in the early 1960s, had just had to write off Superconstellations and DC7s with considerable service left in them because of the arrival of the 707 and DC8. Concorde was dreamed-up purely as a political gesture. On the British side, the Macmillan government saw it as a ticket to enter de Gaulle's Europe. When, on 29 November, 1962, Julian Amery signed the Anglo-French agreement to build the airliner, he did so on the basis of a sketchy, twenty page plan that did not even specify whether the plane was for long or medium range operation. Friends and enemies alike warned the Government that Britain's £75–85 million half share of the cost was bound to be unrealistic. Even *The Times* pointed out that the venture was 'bristling with risks'. But Macmillan and his men, dizzy with Europeanism, went ahead.

They were foiled. The project was barely eleven weeks old when the world's Press was summoned to the Elysée Palace for that historic press conference at which the General said '*Non!*' De Gaulle didn't want Britain in the Common Market. But by skilfully highlighting in his speech the importance of technological collaboration for the future, he managed to thwart any ideas the Macmillan government might have had of trying to withdraw from Concorde.

So the project was on. And the rows and recriminations about its cost began. They reached a crescendo early in 1964, when the Committee of Public Estimates (the House of Commons' financial watchdog) accused the Treasury of being derelict in its duty and criticised the Government for committing Parliament to 'an unspecified heavy expenditure on a project on which the returns must be problematical'. It had taken the intervening nineteen months for Britain and France even to decide what Concorde was for – it was to fly the North Atlantic routes. This had meant a redesign and thus recosting. So it was that Julian Amery, on 9 July that year, announced that Britain's contribution had now risen to 'something like £140 million'. The total bill had doubled already.

On 19 October, after the general election, the new Labour cabinet decided to scrap Concorde. There was dire financial crisis, a run on the pound, and spending priorities to be decided. Concorde had to go. It was either that or leave new school buildings half finished – an unthinkable proposition for a Labour government. Besides, Concorde was a 'Tory prestige project', which could be axed to political advantage. Unfortunately, nobody had bothered to check the original Agreement – which contained no escape clause. The French government was furious, and diplomatic relations were jeopardised. The British Ambassador in Paris learned of the decision to cancel Concorde from a harassed official, hot-foot from the Embassy, while out shooting. Even Roy Jenkins, Britain's new Minister for Aviation (who was not then in the Cabinet), heard the news from a friend at his club in London. Least of all were the British manufacturers – BAC and their subcontractors – properly informed. All hell broke loose. It ended when the British government, not daring to risk the humiliation of an international law suit, back-tracked. Britain cancelled the TSR2 instead.

But two years later, with costs rising again, to beyond £500 million, Mr Wilson's government was still keen to find an excuse to persuade the French to cancel the project. And it was then that the question of noise (both sonic boom and engine noise) suddenly loomed large. Concorde had become a symbol of environmental horror on the grand scale, and various tests were organised with supersonic fighter planes to assess the effects of sonic bangs on the temper of the British people and the fabric of town and country. There was some public opposition, but much less than expected – which probably disappointed the Wilson cabinet.

By now, Concorde was becoming a genuine political football, popular one moment, unpopular the next. In January 1967, Harold Wilson and George Brown went on their famous European tour to discuss Britain's renewed Common Market application, and Concorde was once again a proud symbol of modern science and technology. In Paris, Wilson stressed 'the importance of technology in our approach', and later wrote that 'both of us attached the greatest importance to our bilateral collaboration, defence and civil'. Paradoxically, Concorde became even more secure immediately after devaluation the following spring – despite enormous pressures on the British government to save money. At that time of lost self-confidence and national prestige, a prestigious British scoop in high technology (even a shared scoop) could not be jettisoned.

In America too, politics played a central part in decision-making about whether to enter the field of supersonic transport aircraft. In June 1963, Pan-Am announced that it had taken an option on six Concordes. This was intended to force President Kennedy to back a more lavish American SST programme, yielding fatter profits for Pan-Am. It succeeded. But because a project on such a scale requires massive government support, it threatened the very foundations of American capitalism, and was always bedevilled by accusations of 'nationalisation'. Further troubles included the failure of Boeing's attempted design and the increasing identification of SST not only with environmental pollution, but also with the hateful 'military-industrial complex' in the USA. Eventually, early in 1971, the Senate killed the whole venture.

Another scientific project which has served as a very battered football in the American political arena in recent years is cancer research. During 1970 and 1971, President Nixon and Senator Edward Kennedy were engaged in an escalating battle to see which of them could take the most spectacular initiative in launching a major attack on cancer. It began when Kennedy launched a plan to remove the National Cancer Institute from the National Institutes of Health, Bethesda, and turn it into a mission-orientated Cancer Research Authority, with all the dedication and glamour of a Space Agency. Nixon, responding to the Kennedy initiative, decided to inject massive funds (100 million dollars) into a new cancer research campaign within the existing structure of the NIH. Political support for the Kennedy scheme grew when a nationally syndicated columnist urged her readers to send letters to Congress on behalf of his plan, which produced some 7000 pieces of mail. Nixon then shifted his ground slightly, arguing that cancer research should remain within the NIH but have a special autonomous position. Shortly afterwards, in July 1971, a compromise was worked out. A new 'Conquest of Cancer Agency' was set up within the NIH, but with its director reporting directly to the President. Throughout, the entire affair had as much to do with the political ambitions of both men as with the scientific and social aspects of cancer.

In December 1971, the US Senate also took an initiative, heavy with political undertones, against sickle cell anaemia. About one in ten of American Negroes carry a sickle cell gene – the hereditary determinant of a severe disease in which the red blood cells are unable to carry oxygen properly to the tissues of the body. Such individuals do not themselves suffer from sickle cell anaemia, but anyone who inherits the defective gene from both parents develops the crippling disease, which inevitably shortens life. The disease was, however, oddly neglected until, in his annual health message in February 1971, President Nixon announced that an extra five million dollars would be spent on research and treatment. Not to be outdone, Democratic senators mounted a campaign to increase spending still further and in December 1971 they persuaded the Senate to vote some 142 million dollars over three years for research, screening, and treatment. Meanwhile, the Black Panthers had publicised the disease as 'black genocide' and Muhammad Ali had sponsored a screening programme. The Negroes had begun to feel angry about the lack of cash support for research on, and treatment of, sickle cell anaemia, while losing their former sensitivity about a condition that differentiated them from their White countrymen. White politicians thus found they had to take an interest in this disease of Black voters.

So far, we have considered war-mongering, prestige, national economic growth, practical needs, and political machinations as motives for supporting scientific research. But a sizeable proportion of research spending is by private companies, who sponsor science for one reason only – to boost the company's profits and the returns to the shareholder. I now want to look at this spending more closely, from the point of view of both the community and the company.

First, consider the pharmaceutical industry. Throughout the long history of medicine, but with increasing conviction since Paul Ehrlich began his search for 'magic bullets' at the beginning of this century, Man has sought to combat disease by fabricating drugs with special properties. Instead of using substances from plants and other sources which are found, often by accident, to attack particular diseases, we now try to tailor-make drugs to kill bacteria or interfere with bodily processes by specific means. The axis of modern pharmaceutical research is, in fact, this attempt to replace chance by design. But such triumphs are much rarer than research workers would wish. As we have seen, penicillin was discovered by a remarkable series of accidents, rather than by ruthless logical work towards a defined goal. The central difficulty of pharmaceutical research is the incredible complexity of living organisms which, in terms of experimental investigation, can appear as unpredictability and capriciousness. Moreover, as the easier medical battles are won, those remaining require greater and greater ingenuity. For a few years after the development of penicillin, there was a great surge forward in the discovery of further antibiotics. It was simply a matter of screening soil samples from all over the world for other fungi and bacteria which, like *Penicillium*, might produce antibiotics. But gradually this new lode was exhausted, and biochemists and microbiologists had to consider more logical ways of devising new drugs.

The synthetic penicillin story[20] is an excellent example of such a turning point in drug research. When, in 1954, the then chairman of the Beecham group, Henry Lazell, first discussed with Sir

Charles Dodds the possibility of Beecham entering the search for anti-bacterial compounds, the global foray for antibiotic-producing microbes was still in full flight. But inevitably the returns were diminishing and Beecham saw little point in competing in that area. Dodds recommended Ernst Chain as the person to advise what to do. Chain suggested the idea of making semi-synthetic penicillins – taking existing forms of penicillin and changing them chemically to improve their properties. There was a hope, for example, of tailoring new penicillins that would attack bacteria which had become resistant to existing penicillins. Two young Beecham scientists joined Chain at the Istituto Superiore de Sanita in Rome, where he was then working, and within nine months they had produced reasonable quantities of a potential precursor for building novel penicillins. The team returned to custom-built facilities at Brockham Park, Surrey, and set about trying to exploit the new building block. Next, they found a further form of penicillin, which was causing anomalous results in one of their experiments, and it proved to be an even better starting point than their first choice.

Within two years, the new form of penicillin was patented and in commercial production. Soon afterwards, Beecham began to release a stream of different penicillins – Broxil, Celbenin, Penbritin, and others – whose properties were superior to the conventional penicillins then in use. Beecham, late-comers into the antibiotic field, had beaten the world's giants at their own game. The cost of the research was considerable – but worthwhile.[21] Between 1947 and 1957, when Beecham scientists isolated the crucial core of the penicillin molecule, the company's investment in research was about £2½ million. Afterwards, R and D expenditure increased and by 1966 the cumulative total was some £12 million. A scientific staff of about 50 in 1952 rose to 650 in 1968. Beecham were soon selling their new penicillins in 75 different countries, and overseas sales rose from less than £200,000 in 1959–60 to nearly £8¾ million in 1966–67.

All of this has important implications for the way in which pharmaceutical research is organised. In theory, one requires a means of generating a wide variety of theoretical ideas and experimental approaches to a problem, together with the capacity to exploit success anywhere along the front quickly and efficiently.

In turn, this implies that the skills of scientists working at the frontiers of knowledge should be backed up by talent of equal calibre in the later stages of the innovative chain. It also demands a high degree of flexibility. Those in charge must be willing to redeploy resources of manpower and money to take early advantage of a significant success, and to abandon unpromising lines of research ruthlessly at the right time.

Despite the criticisms so often levelled against it, the drug industry in a free-enterprise economy such as Britain or the United States does fulfil these *a priori* criteria remarkably well. Its creativity is borne out by the fact that the overwhelming majority of new drugs emerge from industrial laboratories rather than from the universities or government research establishments, while a country such as Russia, with its monolithic, state-owned pharmaceutical industry, has failed to produce a single major therapeutic breakthrough in all the years since the October revolution. Clearly, the competitive edge, and diversity of scientific approach maintained by commercial secrecy, are effective forces encouraging innovation. Commercial pressures also ensure a continuous reappraisal of work in hand, judicious pruning of dubious projects, and an acceleration of effort in more promising areas.

There is, however, a price to pay for the prolific inventiveness of the drug houses, and one which could and should be curbed by legislation. Firstly, the level of expenditure on advertising is very high indeed – rivalling that on research. The vast amount of direct-mail advertising sent to doctors is one symptom of this lavish expenditure. Secondly, there is unnecessary duplication in marketing drug preparations that scarcely differ from other medicaments sold by the same firm or by competitors. Both tendencies were symbolised in mid 1971, when Roche, using hundreds of thousands of copies of a lavish 52-page book, launched its new tranquilliser Nobrium. The book contained pseudo-scientific statements and graphs, and was widely criticised. There was already a large number of similar drugs on the market. Two of the most widely prescribed of these – Librium and Valium – were also manufactured by Roche. There was no information in *The Book of Nobrium* about the comparative merits of each – the very information required by the doctor. But tranquillisers are extremely profitable. In the United Kingdom in 1971, for example, 20–23 million prescriptions for them were written, involving a total wholesale cost of £12·16 million. And Librium and Valium, in 1971, were nearing the end of their patent life. That is why Nobrium made its appearance at that time.

Paradoxically, with an inexorable process of mergers and takeovers, the drug industry in the West is itself moving slowly towards the massive, monopolistic system. This must impair creativity, particularly at a time when intractable problems demand new ideas, not simply the application of extravagant resources to well-tried patterns of investigation.

Seen on a broad canvas, there is no doubt that science can be a profitable affair for industrial companies. The Beecham synthetic penicillin saga shows this. The subcontractors developing sophisticated bits and pieces for Concorde know it. At its most grotesque, the profit motive lies behind the development and marketing of enzyme detergents, products of science which carry environmental risks but for which there is no genuine need. (Consumer groups throughout the world have shown them to be little more effective than conventional detergents when used in the same way, and while the early skin and respiratory disorders attributed to their use and manufacture have apparently been conquered, they remain extremely impure and potentially dangerous materials.[22])

In 1967, Dr John Hearle estimated that probably more than 50 per cent (by value) of the entire current output of the British textile industry could not have been made 30 years earlier.[23] This was all a result of scientific research. But when we look more closely at the way in which science is translated into cash profit, the situation is less clear. With textiles, for example, at the time Dr Hearle was writing, a chart of industrial research expenditure in Britain showed the textile industry very near the bottom. The explanation of this paradox is that much of the research expenditure which had changed textile technology (and thus boosted profits) was hidden among the fibre-producing activities of the chemical industry, rather than in the textile industry itself. In the USA too, there is no more than 'reasonably persuasive evidence that R and D has an important effect on productivity increase' in industry.[24]

So it is with the vexed question of basic, curiosity-orientated research as the seed-bed from which applied science (and hence industrial wealth) is supposed to flow. An important book published in 1972, *Wealth from Knowledge*,[25] gave little support to this thesis. The book consisted of a detailed study of 84 technological innovations in Britain which won the Queen's Award to Industry for innovation in 1966 and 1967, and strongly suggested that there was hardly any discernible relationship between basic research and the creation of wealth through industrial innovation. As Lord Bowden said when reviewing the book, 'so much seems to depend on luck, on the coincidence between new problems, new ideas, new materials, new markets and new men. Since nobody knows why the industrial revolution began in England when it did, and why it did not happen somewhere else 1000 years ago, it is not surprising that we should be left with this very unsatisfactory solution.'[26] One of the authors of the book, Professor Freddie Jevons, points out that their failure to find more than a handful of direct connections between science and innovation (they included nuclear power and silicones) is the more striking because they deliberately set out to look for them. 'We certainly do not intend to denigrate science; rather, we want to urge recognition of the fact that its value to industry is less direct and overt than has been commonly supposed in the past. If only the mechanism were more clearly understood, there would be a better chance of increasing the benefits, and the justifications for public support of basic science would be greatly strengthened.'

An examination of the reasons why industrial innovations fail provides similar evidence. According to a report[27] published in 1972 on pairs of competing projects, one of which had succeeded and the other failed, innovations do not normally founder because of tricky technical problems that cannot be overcome. They fail for much simpler reasons – such as an inadequate understanding of the market or user's requirements, or the inadequate rank and authority of key managers involved in promoting the new development.

To some extent, the search for direct evidence of immediate cash benefits from basic science is based upon a fallacy. When an industrial company relates a new product or process to the science behind it, how much research should be taken into account? Research workers can tap the whole corpus of scientific information that has been built up over the years. They will use knowledge that has been acquired, and financed, elsewhere at other times, while keeping their own work secret for reasons of commercial security. In tracing the antecedents of, say, a new piece of computer gadgetry, should one include – and cost – work over the years going back to Babbage and Hollerith? Clearly, that would be absurd. But the problem is a real one. Any R and D department can live off the fat of its own and others' previous research, applying this to practical problems, for a considerable time. There is little doubt, though, that eventually innovation would be impaired. As the Rothschild debate illustrated, the crucial question is *how much* basic and speculative research the community (or a private company) should support at any one time in proportion to science slanted towards particular social, commercial, or other objectives.

What is very clear is that one cannot afford to neglect the social context in translating scientific discovery into applications of practical utility. The case of Polythene illustrates this risk. Research chemists discovered Polythene as the result of a chance observation during research into the effects of high pressure on what are called bimolecular condensation reactions.[28] This happened in 1933 in the laboratories of Imperial Chemical Industries at Northwich. By 1938, one ton was all that had been produced, but the new material found an extensive range of uses during the war as an insulator in telephone and other cables. In 1945, however, Polythene still seemed to be an expensive polymer which would find only specialised application as an insulating material. Plants were built to produce Polythene on this assumption. They proved hopelessly inadequate. Whereas in 1948 just one per cent of Polythene manufactured was for domestic purposes, by 1964 some 40 per cent was used in the home.[29] Between 1943 and 1963, Polythene production in the United States increased from under one million pounds to 2·2 billion pounds, and the price fell from one dollar per pound to 0.13 dollars per pound.

What went wrong with the original estimates? Why, for example, did no one in 1945 think of making kitchen buckets from Polythene? Simply because at that time kitchen buckets were used, among other things, to boil the weekly wash. Polythene melts at 115° C, which is considerably lower than the temperature of a gas flame. It was also expensive, and could not compete with metal buckets. But increasingly even in 1945, the really filthy jobs, which produce dirt that has to be boiled off, were being done by machine. More people were beginning to be able to buy a washing machine, and others were using laundrettes. It was also difficult to forecast the boom in frozen foods, and thus the need for Polythene in packaging, replacing tins. All in all, not enough attention was paid to sociology.

Science in the developing countries

Before leaving a consideration of the motives for supporting science, we must look at the special case of the developing countries. In general, the motives here are as in the developed countries – the hope of gearing science and technology to economic growth, plus the need to solve even more pressing practical problems. Science also has prestige value for a developing country – like having an airline – though this is rapidly becoming tarnished as the ill-effects and abuses of Western science become apparent for all to see. There is also the additional problem of how to retain qualified scientists, and persuade those trained abroad to return home to use their skills. Should the countries of the Third World try to mimic what goes on in the research laboratories in the West? Is there a risk that by ignoring the sophisticated frontiers of molecular biology, for example, and concentrating on short-term pay-off and the eminently practical, they will stifle creative thinking and lose their best brains to the West? Can the whole thing be left to the import of 'technical aid' from the industrialised countries?

These are some of the special questions surrounding the application of science in the Third World. In practice, all too often science is espoused in the wrong way. The crux of the problem was crystallised admirably in an article[30] by Dr Jorge Sabato published in 1970. As a former teacher of metallurgy in universities in Argentina, Brazil, Chile, Mexico, the USA and Britain, Dr Sabato has had wide opportunities to see close at hand the ways in which governments in the Third World – on both the Right and Left wings – can mishandle science. As an example, he cites the three 'life stages' of a public-financed research centre, and a typical government's attitude during these stages.

Stage One, immediately after the government has approved the establishment of a new centre, is a heady, opulent honeymoon. No work gets done, but the money flows like water. The new director is appointed, together with scientific and administrative staff; the building is inaugurated with pomp and circumstance and ministerial speeches; and impressive-looking equipment is installed. The initial investment is very heavy, far outweighing what will be normal operating expenses, and much time and money has to be spent in training qualified staff. But everyone is totally dedicated to the new project, and none less than the government – 'particularly when the buildings are being inaugurated with official ceremonies and patriotic speeches'. No one can possibly complain about shortage of funds. Staff are appointed with a minimum of bureaucratic red tape. And politics do not intervene.

Usually between five and eight years later, Stage Two begins. Scientists at the centre have established contacts with other workers in the same field outside; staff sent abroad for training have returned, fired with enthusiasm and full of good ideas; and the auxiliary technical staff have been trained. Research work now begins to develop, with a slow but sure increase in original work. It is at this point, Dr Sabato wryly observes, that the government loses its enthusiasm for the centre. The honeymoon is over. There are no more functional, modern buildings to inaugurate, and no more impressive equipment to install. Operating costs are now the key item, and the administration becomes thoroughly wrapped around by red tape, and by bizarre restrictions that can make it difficult and complicated to order a new packet of filter paper. Operating costs, as opposed to launching costs, do not yield immediate, spectacular happenings, and funds for vital purposes become more and more difficult to obtain. A request for the appointment of a glass blower can easily cause a bureaucratic nightmare.

As a result, the crucial operational phase slows down. Research becomes intermittent, taking on all the uncertainties of the stop-go tempo beloved of British Chancellors of the Exchequer. Frustration, discouragement, and bitterness spread. A brain drain sets in. 'The centre which seemed so promising now becomes a veritable graveyard for expensive equipment, where only the second-rate staff members remain, continuing to draw their salaries without doing any creative work.' In some cases, Stage Two is arrested suddenly, by a specific political action of the government – often allied with ideological persecution. There are mass dismissals and/or mass resignations, and the creative core of the growing centre is lost. (Again, such nonsense can come from the Right or from the Left.) All too often, therefore, the research centre never attains Stage Three – the time when the centre could be justifying its existence, producing first-rate research work and serving the needs of the community. The bricks and mortar may survive, but the life of the centre is thwarted by bureaucracy and boredom. Even the original purpose of the centre may have been forgotten.

It is a dismal story. But is it any worse than the absurdity of those abandoned research facilities spawned by the *Apollo* programme in the United States? And are we in the West really very much wiser about making sensible use of science? Apparently not. In marked contrast to the speeches of Harold Wilson about white-hot technological revolution, in striking rebuttal of what the philosophers and policymakers of science would have us believe, we are only just beginning to garner the sort of data which will facilitate sound judgments about the administration of scientific research.

Chapter 7
Decisions, pressures, conflicts

One aspect of science and the life of the scientist clashes so violently with the stereotyped image of research as cool, calm objectivity that we often overlook it. Even when we do recognise its existence, we still tend to underrate what is an increasingly important sphere of activity in science. This is the area of public-relations, advertising, grantsmanship, and political machinations. From the individual research worker attempting to loosen the strings of his paymasters, to a science-based company courting public approval for its deeds or a particular scientific lobby seeking to influence government policy, we see increasing evidence of packaging, politics, and persuasion in science and the deployment of science. Indeed, it would be odd if this were not so. Science reflects the social forces which surround it, and we now live in a world where these techniques are of paramount importance.

Take the case of the young research worker in a university department, mulling over a piece of research work he wishes to pursue for the next two or three years. His salary comes from the university, while modest running costs for his laboratory are met out of the departmental budget controlled by his chief. He has formulated his research plans on paper, and concluded that he requires a research assistant (or perhaps a research student or laboratory assistant) for three years, plus an outlay on equipment which is far beyond the financial resources at his command. So he applies for a grant. He may approach a government-financed body, such as the National Science Foundation in the United States or the Science Research Council in Britain. Alternatively, he can seek funds from a commercial company or try one of the big foundations or a charitable institution.

Whichever type of body the supplicant approaches, there are skills distinct from his scientific ability that will help him. He must be able to present his case effectively and do all he can to smooth the way for a favourable decision by the busy committee which will eventually assess his application. Two things help enormously. One is the ability to write a convincing synopsis, in which the proposed project is dressed up in terms both scholarly and seductive. There must be an impressive list of background references, setting the stage for the proposed investigation and signifying the applicant's erudition, and ample recourse to fashionable jargon and concepts of the day. Secondly, even without any hint of corruption, direct personal influence on the committee is of undoubted value. Friends, contacts, lunches can help enormously. The committee is looking to further the progress of science in a way which will also throw credit on the wisdom with which it has distributed the funds at its disposal. But such cabals are invariably packed with currently and previously successful practitioners of the branch of science concerned. They are not, therefore, likely to be receptive to heterodox proposals which offend against their own conventional wisdom. As all the most important advances in science begin with revolutionary ideas, the brighter the theory, the greater the skill needed in presenting it.

As a science writer, one sees the whole gamut of grantsmanship, from inept botching by clever recluses, to dishonest but highly successful operating on a scale that would do credit to the most ruthless American school of salesmanship. At one extreme, there are real live scientists resembling the popular caricature of the eccentric boffin, brilliant of intellect but clueless about the ways of the world. They will complain bitterly about grant applications rejected and frustrations endured in pursuing research with totally inadequate funds. Such people are genuinely unconcerned about money. They believe too that their intellectual gifts are self-evident and do not require packaging. So they are at a considerable disadvantage in a world which now expects even the scientist to have a touch of the salesman and politician about him. Grant-awarding bodies also feel themselves under enough pressure from enthusiastic suitors, without positively seeking out hidden talent. With increasing financial stringency in recent years, this problem has been accentuated, and the lone genius now has a harder time than ever.

Meanwhile, with political and PR skills at a premium, the hustlers and operators at the other end of the scale work even harder. Despite the financial squeeze, cancer research funds have proved a gold-mine for such people. There is a harrowing dilemma in the midst of cancer research which has made such exploitation possible. Cancer is a disorder of growth. Though the fundamental secret of malignancy still eludes us (there may, indeed, be no single cause or secret), all cancers have in common a loss of susceptibility to the controls which normally ensure the harmonious working of the various tissues of the body. In malignant cells, order is replaced by anarchy. This means that virtually any piece of biological research – any investigation, for whatever purpose, on living cells or organisms – could produce results relevant to the riddle of cancer. Indeed, major advances in understanding malignancy have come not only from laboratories devoted specifically to cancer research but from centres where scientists are studying other aspects of living processes.

That being so, it is tempting for a biologist, knowing of the comparatively large amounts of money available for cancer research (much of it from private subscription), to formulate his grant application in such a way as to raise hopes of yielding information relevant to cancer, and send it off to a cancer research body. In its inception at least, the research may have nothing to do with cancer. In his heart of hearts the applicant knows that the chances of relevant spin-off are infinitesimally small. But by skilful presentation, carefully linking his project with cell behaviour and disordered growth, he stands a good chance of receiving funds from a cancer research agency.

The most ingenious piece of special pleading I have ever encountered on the subject of cancer research was in an article by Professor Isaac Asimov in the *New York Times*[1] during the first moon landing in 1969. Noticing the perplexing words 'cancer study' in this article headed 'What samples of the moon may reveal' I found the following chain of reasoning: 'Suppose our analysis of the moon's crust tells us that the chemistry of its organic molecules is at a stage part way to life. . . . We may then have visible clues to the working of the cell. . . . Once we discover the basis in this way, we might have some notion as to what goes wrong in a cancer cell. . . . With that knowledge, biologists might be able to turn to earth cells and, knowing what they are looking for, determine the cause of cancer. . . . This, therefore, is the answer (or, anyway, an answer) to those who ask why we are spending billions to reach the moon, when it is so much more important to cure cancer on earth. All science is one. If we push the boundaries of darkness back in any direction, the added light illuminates all places and not merely the immediate area covered. It is just conceivable, in other words, that by taking the long trip to the moon we will be taking the shortest route to unmasking the riddle of disease on earth.'

Any comment on this harangue would be superfluous.

It is in times of financial stringency that the dangers of autonomy in the scientific community, mentioned in Chapter 3, and those of grantsmanship, intersect forcibly and create particular dangers. The problem is epitomised by the policy of 'selectivity and concentration in research' which Britain's Science Research Council announced early in 1970. Foreshadowed by a speech at Nottingham University on 6 March of that year by its then chairman Sir Brian Flowers, the SRC published a booklet explaining that some research lines were to be singled out for special support, and that in future the SRC would not spread its financial resources as thinly over the ground as in the past. Certain scientific areas, and certain university departments, were to be given 'more favourable than average support', and 'this concentration of resources will be planned by shifting to favoured areas from less favoured areas rather than by simple addition.'[2] As Flowers pointed out at Nottingham, university science budgets were rising at only six per cent per annum in 1970, compared with 12 per cent just five years earlier. *Some* redeployment was thus necessary.

But the principle of selectivity came in for criticism – inevitably largely from individuals and disciplines within the scientific community apparently threatened by the new policy. Professor John S. Anderson, an inorganic chemist at Oxford University, for example, argued that concentration was unavoidable in what is now termed 'big science', where adequate technical facilities could be provided at only a few centres, but that it was less satisfactory for 'little science'.[3] In concentrating resources here, Anderson claimed, the SRC was building on existing strength and fashion, because 'committees are unlikely to operate at the frontiers of knowledge'. It had created bandwaggons – 'glamorous vehicles ... with a marvellous acceleration and very poor brakes. ... Their momentum can carry them on too long.' The risk was that too great an emphasis on limited sectors could freeze the pattern of research a decade hence. 'One has but to look at the journals or the research grant applications, or to interview post-doctoral candidates, to realise how often a line of work outlives its originality and persists, not only through generations of its originator's Ph.D.s, but through his students and his students' students.' The risk is all the greater when a research line is translated into a research department or institute. As times change, some research institutes redeploy relatively painlessly (as Harwell, Britain's atomic energy research station, has contrived to do in recent years, its original purpose largely fulfilled). Many do not.

In the last resort, Professor Anderson argued, neither the progress of science nor profits from science depend on centralised decisions about research projects. 'They depend on backing the right men, irrespective of what they work on.' It was vitally important to 'identify the really bright new stars as they rise – the originators of new ideas and new work which may not fit in with the wisest committee decisions. They may turn up anywhere, and the resources and flexibility to support their work must not be prejudiced by science policy.' It is this capacity to identify the exceptional new scientific brains that is likely to be obscured, particularly when funds are low, by the politicking and grantsmanship of less able research workers. All too often, ability in science is accompanied by deficiency in the other 'skills', and vice versa.

What is particularly galling about the committee system for financing research is that the whole process is conducted with a thin pretence that personal and social factors do not enter into decisions, and that grant applications can be vetted in a state of supreme objectivity, reflecting the austere standards of scientific inquiry itself. It is a sham, encouraged and sustained by the secrecy surrounding the proceedings. Everyone knows this. Members of committees will 'put in a good word' to promote a research grant for a colleague or a colleague of a colleague, in a way they would consider impertinent if such an appeal came from a stranger. Like cabinet ministers arguing over their departmental budgets, they will do deals to help each other. Social contact to ease the strings is considered appropriate by some, not by others – it being more commonly acceptable in the case of a commercial sponsor than with a government or other body.

All of this is not necessarily bad or corrupt. It is wrong that the process is shrouded in secrecy – a body such as the SRC, for example, simply issues lists of grants, without any information about how decisions are reached, and there is no appeals procedure for rejects. It is also bad that there is a double standard of conduct – the theoretical and the real one used by those who know. The situation does indeed raise a wider question – whether the norms of social behaviour common in politics, business and related activities are also appropriate to science.

This question was posed in an important article in *The Guardian* some years ago by Dr Jerry Ravetz, when considering that unique and disastrous project in American science, Mohole.[4] An abbreviation for 'hole down to the Mohorovicic discontinuity', Mohole was a scheme to drill a deep hole through the ocean floor. Its purpose was to facilitate research on the region between the earth's crust and the underlying mantle, in an attempt to learn more about the structure and evolution of the earth. The idea was born one day in March 1957, at a meeting of a panel of the National Science Foundation called to screen some sixty grant applications in the earth sciences. Federal support for science at that time was still booming, but earth scientists felt neglected in contrast to such groups as physicists, chemists, and space researchers. A large and dramatic new project was needed to boost prestige. Professor Harry Hess, chairman of Princeton geology department, later explained what happened at the meeting: 'At the end of a two day session we were rather tired and Walter Munk[5] mentioned that none of these proposals was really fundamental to our understanding of the earth. . . . Walter Munk commented that we should have projects in earth science – geology, geophysics, geochemistry – which would arouse the imagination of the public, and which would attract more young men into our science. It is necessary at times to have a really exciting project. . . . Walter Munk suggested that we drill a hole through the crust of the earth. I took him up and said let's do it.'[6]

Hess recommended that the idea be turned over to the 'American Miscellaneous Society' (AMSOC), an informal and somewhat eccentric group of prominent researchers in the earth sciences set up five years earlier. AMSOC immediately took action and transformed itself into a committee of the National Academy of Sciences, to be the executive agency in receipt of funds from the National Science Foundation. This was a highly unusual move, but not irregular. Those involved were all friends who between them sat on the various official committees dealing with the earth sciences. Shortly afterwards a Russian geophysicist, speaking at a scientific meeting, disclosed that the USSR had plans for a similar project. This greatly eased the flow of money for the American Mohole, and by 1961 small-scale test drills had been completed near La Jolla and Guadalupe Island under an AMSOC project headed by a young, brilliant, but unqualified engineer, Willard Bascom. Early that year Bascom also succeeded in completing a record bore of over 600 ft into the ocean bottom in water about two miles deep.

This was a spectacular achievement, carried through in water ten times deeper than any in which oil drilling had yet succeeded. But it was insignificant compared with the proposed Mohole project itself, which would mean working in three miles of water and drilling about 15,000 feet into the ocean floor. That insuperable problem, still largely unexamined in realistic terms, was one of two disastrous defects in the Mohole project. The other was the failure to think out the goals and expectations for Mohole at the very outset. When things began to go wrong, and as costs began to escalate dizzily, there were bitter recriminations, leading eventually to a scandal of unprecedented scale.

First, the National Academy of Sciences decided to disengage itself. Already, some of the scientists associated with the project suspected they had overreached themselves, and perhaps privately hoped that Mohole would die. But by now the project had acquired an autonomous life of its own, and the National Science Foundation took charge. AMSOC also opted out of its administrative responsibility and took on an advisory role. Confusion followed. While the AMSOC chairman continued to call for an ambitious, multi-stage programme with several different bore holes – still with insufficiently detailed objectives – the NSF invited bids from contractors for a simpler programme using a single hole. With a projected cost at that time of 50 million dollars (the original estimate was five million dollars, and the final estimate 125 million dollars) a row then broke out about the choice of prime contractor. Charges of political favouritism were not quelled when the firm awarded the contract turned out to have close connections with the chairman of the Appropriations Committee of the House of Representatives.

Matters came to a head in 1963, when the Bureau of the Budget froze funds on the project 'until the situation is clarified'. AMSOC members now disagreed about whether to go for one hole or a multiple programme. Then, in the autumn, there were Congressional hearings on Mohole, at which the AMSOC chairman argued for a multistage programme rather than a 'crash Mohole stunt'. The president of the National Academy of Sciences rebuked him for speaking without authorisation, and he resigned at once. AMSOC was later liquidated, and the NSF hatched yet another formula, a compromise between the two types of project. The Bureau of the Budget lifted its ban in January 1964, and the project now moved forwards comparatively smoothly. But it had been fatally harmed. It had lost broad political support, as well as the original enthusiasm of the scientists. Politicians were becoming indifferent to Mohole and in January 1966 the House Independent Offices Subcommittee decided not to allow further funds 'in view of the current world situation'. Enthusiasts made feeble attempts to reverse the decision, but on 24 August that year Congress affirmed its position. Later the same day the NSF announced the end of the project.

'Corruption does not require the offering and taking of bribes by evil and weak men; this is merely its most common and vulgar form,' writes Ravetz. 'Rather it consists in the clandestine adoption of a more lax morality than is appropriate for the social activity being undertaken. The laxity may be in the less severe application of the moral standards of the activity and it may also be in the application of values external to the activity and even contradictory to some of its own.' Mohole was hatched for reasons of publicity evangelism, show-biz – and only secondarily science. It was launched without anyone counting the true cost, without public accountability, and without its advocates properly examining its feasibility or even having a clear idea about its goals. That, Ravetz argues, is tantamount to corruption, and I am inclined to agree with him. Science devours enormous communal resources – of money, manpower, and materials. For that reason, and because, even with greater democratisation of science, it will always be impossible for any one citizen to know in detail what is happening in a staggering range of research specialities, scientists have a severe obligation to explain themselves and not to tap research funds for ambiguous motives. 'When the money flows freely for projects promising prestige and technology, and less so for others, there will always be opportunists around to collect it,' Ravetz concludes. 'And since no planner of scientific work can (or even should) simply scorn prestige or technology, the dividing line between those who are using the publicity machine on behalf of science and those who are using science on behalf of the publicity machine is unavoidably blurred. On occasion, and especially in connection with the biggest projects, the distinction is lost altogether – hence the corruption of science, and Mohole.'

Mohole happened several years ago. It was an unprecedented project in 'big science', and an extreme fiasco involving bungling and serious social wrong. It does, however, contain lessons and warnings which apply to other projects today, large and small. One aspect of the affair distinguishes it from similar wrangles in Britain. Unpleasant though it was, much of the squabbling took place in full public view, rather than in secret. Indeed, the *raison d'être* of the Mohole campaign was to attract public interest, political attention, and thus money, to the earth sciences. In Britain, all too often the scientific establishment tries to have it both ways, courting favourable publicity but with much of the decision-making out of public view.

Such was the case with the row in the late 1960s over whether Britain should join with a plan by CERN (the European Centre for Nuclear Research) to build a new high-powered atom-smashing machine called a 300 GeV proton accelerator – a gargantuan exercise in 'big science'. Such machines are used to probe the nature of sub-atomic matter – the ultimate building blocks out of which everything in the universe is composed. This is an esoteric field of research, scarcely understood even by physicists in adjacent disciplines. At the same time, it is research at the most fundamental level possible, and is thus supported by its protagonists as something of axiomatic importance for the rest of science. But high energy physics is also outrageously expensive. For that reason, outside of the United States and the USSR, the atom-smashing accelerators now required for further 'progress' have to be built as joint projects between different collaborating countries. Dr Francis Cole, a member of the staff of the world's largest particle accelerator at Batavia, Illinois, writing in 1971, pointed out that in the United States alone, high-energy physics research costs something like £80 million per year – 'a substantial enterprise in anybody's terms, but especially so when you consider that there are only a little over 1000 US scientists active in the field, and that their ultimate product is words on paper.'[7]

This is the crux of the argument against such research put forward by many other scientists – even other physicists – who feel that high energy physicists have entered a blind alley where they are dealing, at great financial cost, with artificially-created systems rather than real ones. Further progress will come, such people argue, not from more and more measurements with increasingly powerful machines, but from a revolution of thought leading to a new paradigm of the sort described by Thomas Kuhn (p. 37). In place of the bizarre but apparently meaningless phenomena reported every few months which now compose our confusing and untidy picture of the sub-atomic world, we need a new Einstein to bring order by new intellectual analysis.

Moreover, high energy physics consumes scientific brainpower which could usefully be applied elsewhere. 'At university after university I have seen the more responsible of the elementary particle theorists trying with regrettably little success to put off even brilliant students and send them off to more useful and more rewarding fields,' writes Professor Philip Anderson, a visiting professor at the Cavendish Laboratory, Cambridge, and member of the staff of Bell Telephone Laboratory, New Jersey.[8] 'While some of these bright students are minds that might be equally wasted, from society's point of view, in the more abstract regions of pure mathematics or the like, we must assume that most of them would find otherwise useful employment, especially since so many come from developing countries, where their abilities are of even more importance than in England or the United States.'

Nonetheless, high energy physicists themselves are passionately enthusiastic about their craft and totally convinced that we must go on building the increasingly powerful 'atom-smashing' machines which their work demands. It is not surprising, therefore, that when Britain's decision not to back the CERN project was announced in Geneva on 20 June, 1968, there was considerable anger in the physics community. Yet little of the behind-the-scenes machinations which led to the British government's decision had been made public. We still do not know precisely how this decision – the most important conclusion about the allocation of UK money for science for very many years – came to be made. The most complete account yet available comes from an article by Robin Clarke, then editor of the magazine *Science Journal*, which was published early the following year.[9]

The story began in May 1965, when Professor Denys Wilkinson chaired a committee looking into the requirements of high energy physicists in Britain for the following ten years. After conceding that their own hoped-for target of increased research expenditure over this period was unlikely to be met, the committee calculated what *could* be achieved with annual growth rates of 5, 10, and 15 per cent per annum. Their report was never published, but it provided the basic guidelines on which a decision on the 300 GeV project could be reached. The committee concluded, for example, that with a 5 per cent growth rate Britain would have to choose either to participate in the CERN project and close down its Rutherford High Energy Laboratory or to maintain the Rutherford and back out of the CERN project. At 15 per cent, CERN participation was secure but some sacrifice in the domestic British programme would still be necessary. The Wilkinson report also contained an earlier report by Professor Brian Flowers[10] advocating British participation in the CERN scheme.

In the months following the Wilkinson report British support for the 300 GeV project grew, with articles in the press extolling the benefits of high energy physics. Protagonists set their sights on a CERN council meeting scheduled for December 1967, at which delegates were to be asked if their countries were likely to participate. British scientists convened two further special committees to examine the project in more detail beforehand. Professor Michael Swann chaired one on behalf of the Council for Scientific Policy. It favoured the project unanimously, provided that the overall government vote for science would increase by an average of about nine per cent per annum over the next ten years. The other committee, an SRC committee under Professor Frank Powell, also recommended British participation, but not unanimously. Two distinguished chemists dissented, on the grounds that more than 40 per cent of SRC money would be going to high energy physics if Britain joined the CERN project. It also differed sharply from the Swann report in urging the economic value of building the proposed accelerator at a site in Britain. In a foreword to the two reports, which were published together in January 1968, Patrick Gordon Walker, then Secretary of State for Education and Science, announced that the Government would consider the proposal 'very searchingly' and announce its decision in due course.[11]

In fact, as Professor Flowers (by now chairman of the SRC) learned through 'certain channels', the Cabinet had already discussed the proposal and virtually decided against participation. Because his information was privileged, Flowers was in the tricky position of having to try to influence the Cabinet on behalf of the SRC but without the latter's knowledge. He secured permission to discuss the subject with a few colleagues, and the result was a memorandum to the Cabinet in March in which Flowers suggested that the cost of participation could be met out of the government funds which the SRC was to receive in any case over the next few years, by running down existing accelerators at both the Rutherford Laboratory and at Daresbury. The memo also favoured participation 'on the assumption' that the machine would be constructed in the United Kingdom – suggesting that this might be a bargaining point.

Three months later, at a meeting of the SRC's Nuclear Physics Board, concern was expressed about the delay in the Government's decision, and the Board hatched a plan to salvage the CERN scheme if the government decision went the wrong way. This plan was remarkably similar to that outlined in Flowers's memorandum – although, of course, the members of the Board did not know officially about either the cabinet meeting or the memo. In the event, the Cabinet rejected the memorandum, and Flowers told his Council the entire story at their meeting on 19 June. Several members wanted to resign, on the grounds that the Cabinet, by turning down a scheme for financing the project within the SRC's existing budget, had denied the Council its rightful function of allocating its own money. They were calmed down by the permanent civil servants on the Council, who pointed out that it was not feasible to bargain for a British site (which was unlikely to be selected anyway) and that the run down of the two British laboratories would be administratively extremely difficult.

The CSP was also greatly concerned about the sequence of events, particularly because its members learned formally of the Cabinet's decision only when they read accounts of the CERN meeting held on 20 June. At that meeting Flowers communicated the British government's decision (which he 'deeply regretted') and reaffirmed that the SRC had been prepared to reduce the size of the British nuclear physics community in order to participate in the 300 GeV project.

As Robin Clarke points out in his account of these machinations, one important chapter is still untold – the actual scientific advice on which the Cabinet made its decisions. It looks as though the ultimate influence against the decision came from Sir Solly (now Lord) Zuckerman, who was then Chief Scientific Adviser to the Cabinet Office. The Treasury too was known to be against the project. The Secretary of State for Education and Science was reported to be strongly in favour, though Mintech was against it. The rest is totally obscure. 'What now emerges,' wrote Clarke, 'is that the procedures by which scientists offer advice, and the basis on which major decisions are made, need close examination. And most of all, if any semblance of a democratic process is to be seen to occur, some substantial changes should be instituted. As it is, the most important major decision in science in the UK since World War 2 was taken on unpublished advice which differed from that offered and published by the recognised authorities. It was taken before this advice was presented to Parliament.'

In the event, the whole 300 GeV project stagnated in the months following Britain's withdrawal, with arguments among the willing European countries about money and site. The key to the final solution came from Dr John Adams, who was appointed to re-examine the entire project. He evolved an entirely new, flexible design, allowing the accelerator to be built in stages and increased gradually in power.[12] In December 1970, the British government decided to support this new design and the machine is now being built at Geneva. The distinguished posse of British physicists who had earlier expounded the project in a famous letter to *The Times*[13] ('splendid national heritage . . . enormous advantages . . . obvious benefits . . . cost considerably reduced . . . must be seen to hold our own') were now happy. The new machine, due to start operating in 1976, will cost £110 million.

During the background crisis to the original British decision not to participate, several British science writers were approached in the hope that they might care to write articles emphasising the value of high energy physics and of British involvement in the CERN machine. In general, scientists dislike the Press, but they are also becoming skilful in tapping the services of tame reporters when they can be useful. Scientists' suspicion of the reporter's notebook stems from two principal factors. They have a traditional belief that research findings should not be publicised in the 'lay press' before they have appeared in the 'learned journals',[14] and they worry about their colleagues, who might suspect a yearning for personal publicity in one of their kind found talking to a newspaper reporter. Nonetheless, most scientists can overcome these feelings, particularly if it can be made clear that the reporter approached the scientist, and not vice-versa. (Even as editor of a specifically scientific publication, though a non-specialist one, I have been asked many times to engineer things in this way. One distinguished, retired scientist insisted on starting his relatively innocuous criticism of his former employers with the words: 'I was honoured to accept the editor's invitation to write . . .' though we were desperately short of space for such waffle.)

The role of the press

The business of communicating science has expanded rapidly in recent years. Today there are innumerable press conferences, press releases, briefing sessions, visits, demonstrations, public relations officers, and press officers, to provide the Press with a copious flow of information from the laboratory bench. These things are more firmly established in the United States than in Britain, and are still much commoner on the industrial scene in Britain than in universities. But there is no doubt about the trend –

more information, more rapidly available than ever before; more column inches of laboratory breakthroughs; more, wider, and quicker recognition of scientific achievement.

There are two problems. First, even those scientists who delight in publicity for their intellectual gymnastics are still reluctant to discuss the politics of their trade. Medical researchers in particular are often keen to use publicity where it will help them in attracting cash to their research projects, and will carefully explain the ingenuity of their latest brain child, but become unhappy about discussing the political and social implications of their work. Alvin Weinberg, for example, categorised Dan Greenberg, author of *The Politics of American Science*, as a 'scientific muckraker'.[15] Secondly, it is often the wrong people who exploit these channels of communication the most effectively. The shy biophysicist with great work to his credit hates the very idea of running a press conference. The opportunist biologist, having discovered *the* secret of life (there are at least two dozen such reports every year), rings his favourite reporter, issues a press release for his unfavourite ones, and then expounds his achievement again on an evening television show. Whereas many of the most able scientists are neurotically anxious about what their colleagues will think about *any* press description of their work, others try to exploit reporters as other people use hairdressers and psychoanalysts, to polish the public image and bolster the inner man.

One result of this is that reporters find themselves with the responsibility – which they certainly did not seek – of playing a role similar to that of the referees who serve scientific journals. Every newspaper science correspondent receives a regular flow of both press releases and personal approaches from scientists and PR outfits acting on their behalf, which have to be screened and studied carefully. Many science writers have their mental blacklists. Known cranks apart, these contain people who habitually seek out reporters who will make judicious reference to their work. They will even play one reporter off against another, promising increasingly juicy information for resale to readers. Most science writers have enough knowledge and acumen to see through such tactics. Some have not.

Personal vanity aside, the ability to attract funds by PR is the chief reason why scientists seek to exploit newspaper publicity.

A particularly frank example was recounted in 1971 by a member of an American research group which had been racing neck-and-neck with a team at CERN to discover a new elementary particle, Omega. Talking to the sociologist Gerry Gaston, he gave the following account of what eventually happened:[16]

'Well, finally we found the Omega. We submitted that to the *Physical Review Letters*, and there is a ruling in *PRL* that you're not allowed to publish in any other paper, journal, or anything else, including newspapers, in advance of the publication date of *PRL*. So we called a news conference about a week before, with instructions that the newspapers weren't allowed to publish it. Actually, we got a concession that the *New York Times* could publish in the Sunday edition, in spite of the fact that the *PRL* was to appear on Monday, but they allowed that.

'But there was a leak. The leak occurred through England, oddly enough. Someone in England knew by the grapevine that we had discovered the Omega and wrote a popular article before any experimental results were available. He knew just before publication date in fact that we had done it, and he was tempted – and obviously he fell to the temptation. He did not state that we have discovered it, but he did state that publication was imminent, which it was, but he didn't state publication by whom.

'Then the game goes on. This was picked up by the London *New York Times* correspondent and he knew that this was hot. He then interpreted it as the Omega had been discovered and he didn't know who discovered it and couldn't find out who, but he knew who was looking for it – Brookhaven and CERN. So he said the particle had been discovered in Brookhaven and CERN, but the heading was London. . . . It was a big mess. It killed our publicity. We would have got first page *New York Times* Sunday, which is a very good thing to get. After all, where do we get our national money? From Congress. It makes a great difference if we get first page on the *New York Times* as most congressmen read the front page.

'Then we kicked back. On Friday, we went to *PRL* and said, "They've released it, what can we do now?" They said, "Okay, don't worry" and we backed up the date and told the *New York Times* they could publish. What else could they do? They published on Friday stating really that it was us.'

In other words, the American group wished to publicise their discovery, not to inform the public about work conducted with public money, but to attract the attention of US congressmen. They used an old trick – a press conference with an embargo some days ahead – in an attempt to muzzle the Press until the appropriate moment. They felt resentful when things went wrong, but were still able to exploit the occasion for their much-needed publicity. Not a word, in all of this, about the public interest. It is an unusually honest account. Other scientists indulge in the same sort of tactics, but are loth to speak about it.

On the whole, considering the vast range of subject matter and the sheer volume of information passing through a newspaper office, science correspondents do an excellent job. Even on those occasions when they are criticised for sensationalism,* the net result of news exposure is often positive and good. Such was the case with the world-wide furore over heart transplantation during 1968. Seizing upon the subject for largely emotional reasons after Professor Christiaan Barnard transplanted a heart into Louis Washkansky in December 1967, the Press nonetheless ensured that public attention became focused on the ethical questions surrounding heart transplantation and the unsolved scientific problems which the enthusiasts tended to minimise at that time. Over the ensuing months, surgical teams in various parts of the world tackled the operation – the United States in December and January, India in February, France in April, Britain in May – and in several cases they plainly courted press attention. The British team at the National Heart Hospital in London, in a mood of buoyant nationalism, called a press conference immediately after concluding their operation and sported special ties and 'I'm backing Britain' badges. That was barely five months after the hospital had denied firmly-based press reports that a team there was preparing for heart transplantation, and had issued a statement saying there were 'no plans to attempt such an operation'. It was also, of course, after Professor Barnard had taken the initiative at

* Such charges are much overrated. See *Science* (12 November, 1971, p. 679) for an account by Nicholas Wade of an occasion when careful questioning by journalists forced three prominent US cancer researchers to modify loose and misleading remarks which would otherwise have led to alarmist reports about breast cancer.

the Groote Schuur Hospital in Capetown, and others were moving into the same glamorous field.

Much of the publicity misfired. By triggering off a public debate, it led eventually to exposure of critical questions behind heart transplantation, including those of donors and the 'moment of death'. True, there was emotional and ill-considered criticism. But largely because of exposure in the press, television, and radio, a consensus arose – among both the general public and doctors and scientists initially unfamiliar with the details of the operation – that heart transplantation was ill-advised in the current state of knowledge. Above all, the problem of immunological rejection of a transplanted organ had not been satisfactorily solved. After a flurry of activity throughout the world during 1968, heart transplantation came to a virtual end.

PR and antibiotics—a case history

Publicity is, of course, more often employed to promote the deployment of existing science in the community, than to encourage support for new innovation. One example of this occurred in Britain in late 1969 and early 1970 when attempts were made to influence public and political opinion against the findings of an expert committee which had recommended that the Minister of Agriculture should restrict the use of antibiotics in animal husbandry. The committee had been set up, under Professor Michael Swann, to consider public health dangers arising from the inclusion of antibiotics in animal feedstuffs as prophylactic agents to prevent (rather than treat) disease and to promote the growth of pigs, chickens, and other livestock. The dangers arose from the steep rise in drug resistance in bacteria which had followed the increasing use of antibiotics in farm animals. The bacteria involved were chiefly 'enteric' organisms living in the intestines of animals, many of which can be communicated to man and live in the human intestine too. Dr E. S. Anderson, a bacteriologist working at the Central Public Health Laboratory, Colindale, had first called attention to such dangers in May 1965, when he described strains of the food-poisoning bacterium *Salmonella typhimurium* that had acquired their resistance to ampicillin as a result of the use of this antibiotic in the intensive rearing of calves.[17] Moreover, because some of the resistance originating in this way proved to be of the transferable type – capable of being transferred from one organism to another[18] – there was a possibility that animals could become reservoirs for resistance which would later be transferred to bacteria specifically dangerous to man, such as the typhoid bacillus. Dr Anderson therefore called for 'a re-examination of the whole question of the use of antibiotics and other drugs in the rearing of livestock'.

These warnings were soundly substantiated even in 1965, and were further vindicated in the months following by other research workers throughout the world. Eventually, prolonged public and professional pressure[19] forced the Government to appoint the Swann committee in May 1968. As the committee deliberated, it became clear to those who gave evidence (and was passed through the grapevine to others) that Professor Swann and his colleagues were likely to recommend restrictions on the uses of antibiotics in farming. As a result, a variety of documents appeared in attempts to influence public opinion – and Members of Parliament in particular – against any such restrictions. In September 1969, for example, the Office of Health Economics published a booklet[20] on the subject which did not refer to a single one of Dr Anderson's research papers. The booklet claimed to provide 'a broader economic background to the conclusions of the Swann committee'. In fact, it set out carefully the economic benefits of using antibiotic feed supplements – broadly speaking, quicker rearing and thus cheaper food – without doing justice to the accompanying hazards. The Office of Health Economics is financed by the Association of the British Pharmaceutical Industry, which includes the companies manufacturing antibiotic feed supplements.

Earlier, Crookes Laboratories Ltd, had hired a public relations firm to publicise a series of lavish booklets calling for a 'more responsible approach to the use of antibiotics', but which gently played down the dangers arising from the use of antibiotics as animal food additives. Some months before, the same company had begun promoting to farmers a new 'broiler programme', which entailed giving two antibiotics, erythromycin and chloramphenicol, to all the birds in a flock by adding the drugs to their drinking water. Both antibiotics are used to treat both animal and human infections, and chloramphenicol is still the only antibiotic available for treating human typhoid fever. *Any* increase in resistance to these antibiotics is thus highly undesirable, and Crookes's action was widely criticised by veterinary and other experts.[21] Some months later, the product was taken off the market.

As anticipated, when the Swann report[22] appeared in November 1969, it recommended restrictions. The committee had done its work thoroughly and spelled out its case with clarity. The recommendations were undoubtedly sensible. They were largely adopted by the British government, and later the Food and Drug Administration curbed the agricultural uses of antibiotics in the United States even more severely. But the Swann report in Britain met fierce opposition from the drug industry. This took many forms, from conventional lobbying of MPs by industry spokesmen, to some unhappy behind-the-scenes activity. One example was a symposium on 'the implications of infective drug resistance', held in London on 19 January, 1970 by Cyanamid of Great Britain Ltd – the English subsidiary of the American parent company – in a transparent attempt to influence the debate then raging over the report. Neither invitations to the symposium nor the programmes carried the sponsor's name. But they did include the name of the Royal Society of Medicine – the venue booked for the occasion by Cyanamid – in such a way that invitees might well have concluded that the RSM had sponsored the meeting. The RSM, a prestigious learned society which holds regular meetings of its own, became aware of this critical omission only when an unwanted invitation was returned to them. The symposium itself, with three platform speakers arguing strongly against the type of controls recommended by the Swann committee and one neutral, confirmed suspicions that Cyanamid were more concerned to mount a one-sided onslaught than to catalyse a real debate among the various protagonists.

The day following the Cyanamid symposium, there was a press conference at the Carlton Tower Hotel, London, at which journalists were invited to hear a 'case for a balanced scientific approach by governments to the control of human, animal, and plant health products'. The speakers were Dr Robert White-Stevens and Dr Thomas Jukes – two of the panel from the previous day's symposium. Both were concerned, the invitation said, about 'instant decisions' by governments on such substances as antibiotics in animal feedstuffs, without full discussion of the risks involved. The conference was organised by a public relations firm (the Graham Cherry Organisation Ltd), but the invitations gave no client's name. At about the same time, this organisation issued a postal survey of the Swann committee proposals, conducted among farmers and 'other parties most concerned with the recommendations' (which did not include microbial geneticists, the people with the greatest understanding of the subject and therefore of the risks involved) and commissioned by 'leading pharmaceutical firms'. They also issued an 'economic appraisal' of the proposals, carried out 'on behalf of a group of pharmaceutical concerns'.

There was an interesting repercussion to these events. As editor of the weekly magazine *New Scientist*, I published a factual account, and made restrained criticisms of anonymous and semi-anonymous sponsorship in scientific discussion, particularly in an area with heavy commercial involvement and far-reaching implications for public health. I was widely supported, and the Royal Society of Medicine later tightened-up its instructions to companies hiring its premises to ensure that organisers' names were included in all literature prepared for such meetings. Even the Institute of Public Relations made a new ruling that names of sponsoring firms must be printed on invitations to press conferences and the like. The various parties had their say in the correspondence columns of *New Scientist*. But this was not enough for Cyanamid. The company's managing director wrote 'without prejudice' to the chairman of the International Publishing Corporation (owners of *New Scientist*) drawing attention to the 'very grave charges indeed' and 'damaging allegations' which we had made, pleading innocence, and clearly expecting the chairman to take internal action. The chairman supported me, and that was the end of the matter.

Chapter 8
Science under attack

'I hate and fear "science" because of my conviction that, for long to come if not for ever, it will be the remorseless enemy of mankind. I see it destroying all simplicity and gentleness of life, all the beauty of the world; I see it restoring barbarism under the mask of civilisation; I see it darkening men's minds and hardening their hearts; I see it bringing a time of vast conflicts, which will pale into insignificance "the thousand wars of old" and, as likely as not, will whelm all the laborious advances of mankind in blood drenched chaos.'

George Gissing, writing at the turn of the century,[1] foreshadows in those words some of the major developments for which science and technology have been so fiercely pilloried in recent years – the destruction of life and beauty by pollution, the harnessing of science to military destruction, the blatant technology exploited with cavalier regard for its effects on people and society. These are all potent factors behind the various movements – from the philosophical to the eminently practical – recently deployed against science and technology. They have gained additional confidence with the increasing growth of consumer movements and direct action politics during recent years. So it was, for example, that the first issue of *Undercurrents*, an 'underground' science publication launched in London in 1971, contained not only a leader on the theme that 'technology no longer concerns itself with the satisfaction of human needs, but with the churning-out of cheaper and ever-more sophisticated products which the masses can be persuaded they need', but also practical technical advice about setting up small community-based radio stations.

In this chapter, I shall examine a number of different movements and influences working recently and currently against science and technology in their various guises, which have been mounted from outside the scientific community. Critical movements within the ranks of professional scientists will be covered in Chapter 10. There is not, of course, always a rigid separation between the two. When, for example, a campaign was fought in 1964–67 over a proposal to build a new reservoir at Cow Green in Teesdale, to provide water required by Imperial Chemical Industries at Billingham,[2] the forces mobilised against the reservoir included an alliance of professional botanists anxious to preserve a unique ecological site, amenity societies, and private individuals concerned about the destruction of an attractive part of England. Similarly, the anti-Concorde Project in Britain has a mixed committee and membership of experts and non-experts. In most cases, however, it is possible to see clearly from where the initiative comes, and in this chapter we shall concentrate on criticisms originating from outside the scientific establishment.

First, some indications of the strength of feeling displayed recently on a range of scientific and technological issues affecting the public. In February 1971, the Dutch government was almost compelled to postpone a census – a complex operation involving 100,000 census takers and 110 million cards – following public and parliamentary pressure stemming from fears of computers and data processing.[3] A public opinion poll just before the census showed that a substantial minority of the Dutch people were unhappy about it and would refuse to co-operate. Suspicion of computers was the chief cause of people's distrust of the census, according to Ch. Fransen, statistics bureau chief for the Rotterdam municipality. 'Merely the word computer provokes anxiety and alarm in many people' he was reported as saying.

Similar indications came from the prolific flow of letters and inquiries received by Mr Harvey Matusow, when he brought his International Society for the Abolition of Data Processing Machines to Britain in 1969. Every time Matusow was mentioned in the press or on the air, correspondence flooded in – 350 letters following a short item in the gossip column of *The Observer*, 600 more after a four minute item on a radio disc show at 12.25 a.m.[4] Inquiries and support came from activists willing to indulge in anti-computer warfare, professional scientists and technicians, clergymen, businessmen – a wide range of sympathisers, suggesting a genuine groundswell of antagonism towards the spread of computing and data processing. Some letters were from victims of computer errors – the miscalculated bank payments and similar horrors. Some were blatantly luddite in tone. Many were more positive, and were from computer professionals who nonetheless felt strongly about impairment of confidentiality and other possible abuses of the computer.

A third, entirely different, episode: Towards the end of 1970, following a BBC-television programme *Go Climb a Mountain*, there was considerable public interest in Britain in the allegedly high cancer-cure rates being achieved by Dr Josef Issels at his Ringberg-Klinik in Bavaria. The medical basis of Issels' treatment was controversial. While specialists in Britain generally agreed that his personal approach to his patients and high standards of medical and nursing care were beneficial, certain of his measures designed to stimulate the body's own capacity to reject cancer were felt to be dubious. Because Issels did not keep patients' records in the conventional style, it was difficult, if not impossible, to compare patients treated by Issels and similar series treated differently elsewhere. On the other hand, there was at least enough *prima facie* evidence to justify optimism. Public anxiety grew, and eventually this pressure led to a high-powered team of British cancer research specialists making the trip to the Ringberg-Klinik to study Issels's methods and results.

The *British Medical Journal*[5] was very angry: 'To carry out their study, these exceedingly busy people had to interrupt their work in the laboratory and at the bedside for some weeks . . . Popular clamour for a will-o'-the-wisp has occupied the time of people who are much more likely to bring benefits to sufferers from cancer if left to get on with their work. In a microcosm, this whole unhappy episode exemplified how not to influence research . . . Pressures of the kind to which the Cancer Co-ordinating Committee felt obliged to respond can severely harm the interests of the people applying it.' In other words, the laity must not trouble busy doctors and medical scientists. They must accept complacent assurances that all is being done for the best. As Clifford Brewer, senior consultant surgeon at Liverpool Royal Infirmary, said in a letter to *The Times*[6] just after the controversy broke: 'let it be said that the patient with advanced malignant disease is more or less guaranteed a dignified form of death with continued help under the state service'. In the event, although the committee found no conclusive evidence of above-average cure rates, they did learn important lessons about patient care and measures to boost morale, which have since filtered gradually into use in Britain. The visit was vindicated.

Not one of these episodes would have occurred even ten years ago. Then, people still believed what doctors told them, took part dutifully in computerised censuses, and did not feel moved to protest about the dumping of toxic chemicals. Other types of movement dedicated to criticising particular aspects of science have a longer history. It is to these, which have grown in strength in recent years, that we now turn. They consist of the heretics, the cranks, and the outsiders who are usually dismissed with irritation or patronising complacency by the scientific establishment. They are, in fact, vitally important to the health of our society, in three principal ways.

First, there is the question of native wit, the gut feeling that is so often reliable while flying in the face of conventional wisdom. It is this gut feeling that nourishes many dissident groups, which in turn stimulate and support the feeling in outsiders, attracting them into the fold. An excellent exponent is Britain's Soil Association. Since it was first launched in 1947, the Soil Association has drawn its lifeblood from just such a source of intuitive judgment – a quasi-mystical affinity with the organic processes of nature. Dismissed as eccentrics, and sometimes ridiculed by agricultural scientists, its spokesmen were warning the public about despoliation of the countryside, environmental rape, and the evil side-effects of modern technology long before these topics became fashionable fodder for after-dinner speakers. Academic agronomists listened politely to Soil Association warnings that modern methods of husbandry – particularly the intensive use of inorganic fertilisers and heavy machinery – were ruining the structure of the soil, but said that there was no cause for worry. Experts from all sides were happy to join in the critical chorus. But the Soil Association men were right. Gradually, it became clear that soil in many parts of Britain *was* being broken down by modern farming methods. An Agricultural Advisory Council survey, carried out in 1970 following growing concern and pressure on the Ministry of Agriculture to do something, showed that the soil in many areas of England and Wales was being chronically damaged.

The second worthy and tenacious quality of the outsiders is their staggering capacity to gather information from all corners of the globe to support their case. The anti-fluoridationists, campaigning against the addition of fluoride to public water supplies, illustrate this magpie-like dedication particularly well. They will seek out every new research paper, every qualified opinion, every survey report, however obscurely published, and add it to the dossier used to bolster their case. Such avid screening is as efficient as a systematic information retrieval system in ensuring that all relevant information is gleaned and scrutinised. This is the way, for example, that the National Anti-Fluoridation Campaign works in Britain to attack what they call the 'government sponsored hoax' of fluoridation. Given the ease with which a plausible case can so quickly become unquestioned orthodoxy, and be accepted uncritically, this is an invaluable activity.

Perhaps the most important feature of dissidents, however, is their stimulatory role. Simply by preaching loudly enough, for long enough, they help to mobilise public and professional opinion and expose weak points in conventional practices. This is where the anti-vivisectionists are so useful. Despite the shaky foundations of the extreme anti-vivisection case, such pressure is useful in adding impetus to moves to replace animals by non-animal techniques in medical research. It can be argued that the high cost of laboratory animals alone forces research workers to seek alternative techniques. But against this there is the influence of fashion and familiarity, which leads people to retain well tried and tested methods. As with factory farming, constant reminders from the anti-vivisectionists – including discomforting publicity – can usefully reinforce other pressures against animal experimentation, as well as providing an extra check on anyone tempted to transgress codes of conduct. In recent years, moreover, the anti-vivisection movement[7] has turned away from outright opposition to all animal experiments and the use of exaggerated figures to present its case. Instead, anti-vivisectionists have begun to exert greater pressure on research workers to devise alternative techniques. That is the basis of such organisations as the International Association against Painful Experiments on Animals, launched in 1969, and FRAME (Fund for the Replacement of Animals in Medical Experiments), set up in Britain in 1970.

From the examples we have considered so far, one characteristic has already emerged. Movements which appear, on the surface, to be tiresomely negative, do in fact produce positive results. The pressure expressed and mobilised, although directed *against* some sector of science or the application of science, nonetheless leads to *positive* improvement. Such is the case in matters of ecology, pollution, and the environment. Largely as a result of unprecedented public concern and pressure, there have been very real improvements in the environment in recent years. The oft-quoted but dramatic improvement in the condition of the river Thames is a typical example. In 1958, the lower Thames was so heavily polluted that it was usual to find no dissolved oxygen whatever in the water for several months of the year. The river stank, and there were no fish. Today, the water contains oxygen throughout the year, the smell has gone, and fishes have returned.[8] On a larger canvas, all Western governments, pressured into acute awareness of environmental problems, have passed a steady stream of environmental legislation designed to reduce air and water pollution, combat public nuisance, and maintain and improve amenity. The amount of sulphur dioxide in the air of sixty American cities fell by a third between 1962 and 1969, while smoke in the air declined by 20 per cent between 1957 and 1970. Political parties now vie with each other to promote the cause of environmental cleanliness. Industrial companies are under similar pressures to be seen to take responsible measures to curb pollution. At the global level, governments too have accelerated their efforts towards international agreements to facilitate pollution control. The mammoth UN conference on the environment, held in Stockholm in June 1972, was one major symptom of this new mood.

Without question, much remains to be done. What *has* been achieved so far has been to a large degree a result of an astonishing flourishing of public concern over the past ten years or so. Many groups, organisations and publications have contributed to and channelled this concern.[9] Rachel Carson wrote the right book, *Silent Spring*,[10] at the right time. In the United States the magazine *Environment*, and in Britain *Your Environment* and *The Ecologist*, have concentrated upon the fight for a better environment. Even a Bishop, Hugh Montefiore, wrote a book on the environment.[11] And in November 1971, pollution was heralded as 'a new dimension in the political struggle of the working class' at a meeting of the Gramsci Institute of Communist Studies, held in Rome. Groups such as Friends of the Earth, launched in the United States in 1968 and Britain in 1970, have done excellent work in publicising such issues as whale conservation and the need for more recycling of materials. They also published excellently detailed profiles of the 1972 US presidential candidates from the conservationist standpoint, emphasising their records, achievements and deficiencies in environmental affairs.

The approach and flavour of the Friends of the Earth can be gauged from these extracts from a newsletter by their London group, published in the April 1972 issue of *The Ecologist*:

'It is Monday, 14 February, 1972, the sun is shining and 123 unopened letters await the arrival of the three Friends who will spend most of their day attempting to answer them . . .

'Simon Millar is looking up the latest figures on known world coal reserves. At the present rate of usage coal could last 900 years: with the current small rate of increase of the amount used each year, coal will last nearly 600 years. So that's not bad – or is it? What's going to happen to the rate we use coal when we run out of other fuels?

'The sun is still shining, but with the office full of people answering mail, planning campaigns, catching up on research work and chatting about energy supply, no one has mentioned nuclear power. Strange that. A year ago, Simon, Tina, Angela or Jon would have chipped in something about nuclear energy. They're quiet about it now not because of the research they've done into radiation hazards (although that's enough to worry anyone) but because pinned up on a wall is the second Law of Thermodynamics: "All energy eventually ends up as heat." The trouble with that law is that it is not man-made, and can't be broken. So with energy the problem is not just one of running out of "safe" fuels like petroleum, but also one of not knowing how long we can carry on doubling our energy use without producing enough waste heat to upset world climate . . .

'The mail's almost done now, and Simon, Angela and Peter are preparing three different campaign schedules. By the time this . . . is read, things will have moved on quite a bit in each of FOE's three main campaigns. On 24 February is . . . National Parks Day, and Simon, who prepared most of our mining evidence to the Zuckerman Commission, is painstakingly sending notices and posters to the headquarters of the 30-odd organisations holding public meetings on that date. That's Friends of the Earth: 50 per cent academic research; 49 per cent stamp-licking; 1 per cent excitement. If only people knew . . .

'Angela's arranged a deputation to go to the Department of Trade and Industry on 1 March to argue the toss about Endangered Species Protection. We've done the homework, prepared the papers, organised the Adopt-a-Species Campaign, and now it's all up to Anthony Grant and the DTI. If they care about snow leopards and blue whales (an awful lot of politicians pretend to – when it suits them), then the government will agree to introduce an Endangered Species (Importation of Products) Bill, banning the importation of products made from these animals. If you haven't heard what the government intends to do, ask Anthony Grant, Under Secretary of State, DTI, 1 Victoria Street, London SW. Send copies of your letter and reply to us at 9 Poland St, London W1 . . .

'Peter Wilkinson is getting ready for the big one. On 25 March over 500 groups up and down the country will be taking back whatever excess packaging they have collected in the last month. Administratively, Pete's task is nigh-on impossible, but then so was the National Schweppes Delivery last October, and he organised that.'

Let us turn now to some specific examples of environmental action. First, the uproar in the United States in the late 1960s over the siting of nuclear power stations.[12] Opposition to projected reactors began in earnest in 1967–68, on the basis of radiation hazards. There were fears of both routine discharge of dangerous amounts of radioactive waste materials, and of massive radioactive hazards in the event of accidental breakdowns. In general, the nuclear power industry was able to discredit these views by arguing that the environmentalists' estimates of the risks were exaggerated. Attention then moved to the question of 'thermal pollution' – the outpouring of warm waters into rivers, which can cause adverse ecological effects. Again the Atomic Energy Commission refused to budge, this time on the grounds that its concern was radiation safety, not hot water. The opponents then returned to radiation risks and gradually mounted a more and more impressive, detailed case against certain types of reactors and discharge limits. Slowly, it emerged that there *were* no black and white answers (see p. 208). The environmentalists were *not* as foolish as many of 'the experts' had tried to make out. At last, the public became aware of the real problems behind the design and operation of nuclear reactors. And US action has precipitated similar concern and campaigns in many other countries.[13]

Again, the end result – after much irritation for nuclear engineers – was better protection for the citizen. We have this on the authority of Alvin Weinberg, director of Oak Ridge National Laboratory, speaking at a Ciba Foundation symposium on 'Civilisation and Science', held in London in June, 1971. Following continuous public exposure of the subject, he says, 'reactors now, at least in the United States, are loaded down with safety system after safety system – the safety and emergency systems dominate the whole technology. By contrast, in the Soviet Union, where the public does not have automatic right of access to scientific and technical debate of this sort, the technology of reactors is less obviously centred around safety. Soviet reactors until recently had no containment shells, no emergency core cooling systems, no pressure suppression systems . . . One cannot attribute these differences simply to the existence of the very influential Advisory Committee on Reactor Safety in the United States, since review mechanisms exist in the Soviet Union. I would rather attribute the contrast to the difference in the degree of access of the public to the technological debate.'[14]

Another intriguing area where public concern has had a real, though undramatic, effect on environmental science is in the development of pesticides. Following the publication of *Silent Spring* in the United States in 1962 and Britain the following year, there was public outcry in both countries about wildlife deaths caused by Aldrin, Dieldrin and other insecticides. One result was a series of bans and partial bans in different countries on the use of such compounds. Another, revealed some years later by an analysis of the entomology literature, was a decline in the number of research papers published on chemical control of pests, and a corresponding increase in research reports on ecologically-based control.[15] It seems that – at the very least – this change would have occurred more slowly without public concern and a consequential political campaign against chemical insecticides.

Altogether more straightforward than the intricacies of pest control or nuclear power generation was the case of action against the United Carbon Black factory in Port Tennant, Swansea, in 1970 and 1971. In January 1970, a band of housewives living near the carbon black factory dumped their dirty washing – blackened by filthy smuts from the factory chimney – at the Guildhall in Swansea; demanded measures to curb the outpour of filth; and returned home to block the road outside the factory. In response, the factory management installed new equipment which was supposed to reduce the amount of smut. It did not work. Neither did renewed residents' protests to the Council, Members of Parliament, or the local health department. At a meeting on 26 January, 1971, the people affected decided to block the road leading to the factory permanently until the filth problem was tackled properly.

On 1 February, it was announced at a Council meeting that the factory was planning to increase its production by 25 per cent. At 9.30 a.m. on 3 February, fifty housewives moved on to the road outside the factory, determined to maintain a blockade until they achieved their objective. They pitched a tent on the road, and brought furniture, food, radios and other equipment. Local tradesmen delivered regular supplies of coal, and other goods. Using shifts of fifty people at a time, the vigil soon settled down as a day-by-day-activity. The whole pattern of life changed to account for 'blockade duty'. Despite bad weather, morale remained high, and by the end of three weeks, three departments had to be closed down, as products could not leave the factory or raw materials arrive.

Two reports were published, one by an Alkali Inspector from the Welsh Office, another by Britain's Deputy Chief Alkali Inspector. Both declared that the carbon black factory was meeting its legal requirements. But the people refused to move. Production was now seriously affected. On 26 February, the road blockers met the factory management and Swansea Council. The result – a famous victory for the protesters. The management promised to spend £200,000 on pollution control. The factory would be thoroughly spring cleaned, lorries would be re-routed to minimise nuisance, and production would cease whenever strong easterly winds were prevalent. A liaison committee, including residents' representatives, was proposed and there were hints that the company might drop its ideas for expansion.

The residents lifted their blockade. Three days later, the company announced that it was to shelve the factory extension plans. As reported in the activist publication *Solidarity Swansea*, 'the people of Port Tennant had established some important principles, and shattered some myths in the process. The management of a large factory had been forced to allow those who lived near it to have some measure of control over its production . . . Direct action has gone beyond the range of the symbolic protest: you don't show that you could close the factory if you wanted to – you try and do it!'

The Port Tennant episode was a relatively clear one for those involved on both sides. Many other clashes between citizens and science-based industry or modern technology are more complex. Their resolution often depends upon considerable legal and scientific expertise and information. This is costly, and if those fighting lack the necessary resources the dispute can be an absurdly one-sided affair. At the public inquiry which preceded the building of a new aluminium reduction plant near Holyhead in Anglesey in 1968 by Rio Tinto-Zinc, the Kaiser Chemical Corporation, and a consortium of other companies, opposition came from a handful of small bodies including a small residents' association and the British Lichen Society. Against the vast financial and expert resources of the consortium, there was no chance whatever of the opposition case being presented equally effectively. The company was given nine days to argue its case, and the objectors two days.

Contrast this with the successful campaign mounted to persuade the British government not to build its third London airport at Cublington in the Buckinghamshire countryside – as recommended by the Roskill commission set up by the Government to compare different sites. Money and public relations skill were crucial. At an early stage the Wing Airport Resistance Association was created to fight against the selection of Cublington and had soon attracted 62,000 members. The hard core committee were from the law, accountancy, advertising, and other professional jobs, and included Lady Hartwell, wife of the chairman of the *Daily Telegraph*. With skill and cunning, they first set out to persuade the Roskill commission to reject Cublington. But by June 1970, having spent £50,000 on legal fees to counsel, the Association realised that this battle was lost. Roskill was likely to select Cublington. So the Association found more money, hired one of the biggest PR agencies in Britain (Burson Marsteller), and began trying to influence Parliament instead. Months of intense and skilful lobbying followed. The end result: Roskill came down in favour of Cublington, but the Government rejected its advice and decided to build the airport at Foulness on the Thames estuary. An anti-Foulness committee raised about £3000 for its own protest movement. The Cublington group spent at least £700,000. The leader of the Foulness group said: 'The Cublington lobby's ready access to government, MPs, press, television, and radio has been one of the most depressing features of the whole operation.' It was, in fact, crucial.

Can the process of critical attack on science and technology go too far? Many people feel so. In particular, there is a continual danger that fashion may play a capricious role in the selection of targets for dissent. In 1961, for example, Consolidated Edison, which supplies power to New York City, announced plans to build a nuclear power station on the East River. The proposal was quashed after a public hearing at which environmentalists emphasised the risk of radiation. The company built the oil-fired Big Allis generator instead. But the East River already had some of the worst polluted, most breathed, air in the world. Big Allis simply added to it. Public concern over atmospheric pollution mounted, and in 1966 the city signed an agreement with Con Ed that no more fossil fuel plants would be built in the city. By implication, nuclear stations would be allowed, and advertisements in the press began to emphasise the cleanliness of nuclear power. By 1971, however, the pendulum had swung again, and in April of that year, stimulated by public protests, the Atomic Safety and Licencing Board refused permission for Columbia university to fuel and operate a Triga reactor to be used for research. The reactor – one of the safest designs ever – had been built, but could not be operated.[16]

There have been fears, too, that in the matter of drugs and food additives, the critics have been *too* effective, pushing governments into excessive legislation and control agencies into panic decisions. In the United States, Senator Estes Kefauver led the way with his series of hearings in the late 1950s on questions of drug pricing and safety. Though they received little attention in the early days, they became the focus for considerable public attention following the thalidomide tragedy of 1961, which precipitated an outcry against the drug industry on an international scale.[17] Since then, agencies such as the Food and Drug Administration in the USA, badgered by public and political pressure, have imposed increasingly exacting standards for new drugs, including extensive testing to reveal rare side effects and long-term toxicity. Kefauver's role was later taken over by Senator Gaylord Nelson, who has played a unique role as a consumer advocate in Washington, largely in the area of pharmaceuticals and food additives. In July 1971, after investigating the drug industry for four years, Senator Nelson's subcommittee had generated 21 fat volumes of hearing testimony, most of it denigrating the industry, the FDA, and doctors.[18]

It is, of course, right and proper that bodies such as the FDA exist to protect the public. But in recent years there have been signs, particularly in the USA, that the authorities may have gone too far in some instances, forgetting the need for a balanced appraisal of any untoward side-effects alongside the benefits of particular drugs. A number of well-established remedies, if introduced for the first time today, would not survive the scrutiny that is necessary for new drugs. There is also the risk that apparently unfavourable evidence about a drug or food additive may be accepted over-hastily as a result of public or political pressure. An obvious example was the ban, in both the United States and Britain, of the sweetening agent sodium cyclamate, in panic circumstances in October 1969. The evidence on which cyclamate was proscribed was ambiguous, and was based on experiments in which rats developed tumours after consuming gargantuan amounts of cyclamates – 50 times the maximum recommended daily human consumption for their entire lifespan. Other experiments showed that implanting pellets of cyclamate in mouse bladders caused tumours. Using such techniques, however, it is possible to throw suspicion on any substance, however innocuous. Even common salt, when injected in large amounts in a particular way, can cause cancer.

One result of sustained public pressure against 'chemical contamination' was illustrated by the reaction of some pesticide manufacturers in the USA in the late 1960s and early 70s. Tired of continuous public attacks and legislative pressures, companies began to cut back on their investment in pesticide research, to such an extent that entomologists expressed fears about a possible 'pesticide vacuum', with a gradual drying-up of the flow of new products. Similar reactions followed the banning of the contraceptive 'minipill' in January 1970. This type of pill was developed specifically as a means towards safer contraception. The combination of its tiny dose, coupled with the elimination of oestrogen (the component believed to be implicated in the minutely increased risk of thrombosis in women using the conventional pill), pointed to the minipill being completely free of side-effects. Seven years of research on a drug which had been studied in greater detail than any other, had failed to produce any evidence of danger. But, again on the evidence of some ambiguous and dubious results – the finding of breast nodules in bitches after fourteen months' exposure – and under pressure from Senator Nelson's sub-committee, the FDA banned the minipill. Britain followed suit a few days later. Not long afterwards, one company in Britain which had pioneered fertility control announced that it would phase out its research in this area.

Surprisingly, America's pioneer in aggressive consumer research, Ralph Nader, has paid little attention to the pharmaceutical industry. The product of modern technology to which he has devoted the closest attention is the automobile – safety standards and pollution in particular. But it was Nader who inspired James Turner, leader of one of Nader's study groups, to write his widely publicised book *The Chemical Feast*,[19] published in 1970, which attacked the food industry, largely on the inclusion in modern foods of a whole range of chemical additives – flavours, colouring agents, extenders, 'improvers', emulsifiers, aerating agents and others. Again, this type of scrutineering exercise is useful, but can go too far. No food is completely safe. Potatoes contain about 90 parts per million of solanin, of which 400 parts per million have been shown to be poisonous. An ounce of cress contains enough benzyl thioglucoside, producing enough benzyl cyanide, to kill two mice. Cabbage, charcoal-grilled meat, lettuce, kippers and many other foods carry measurable amounts of carcinogenic compounds.[20] Where, then, should we draw the line? Are 'natural' foods to be considered more leniently than newly proposed food additives?

Two important conclusions emerge from the various cases of criticism and dissent we have been considering. First – whatever the nuisance they feel to be caused by protesters, critics and activists – professional scientists and technologists have little choice but to expect *more* public involvement in scientific and technical matters. Moreover, it will become an increasingly active involvement, and one dependent upon using the same weapons employed by the experts. Blocking the entrance to a carbon black factory is a relatively crude, but effective, measure to combat a relatively crude nuisance caused by technology. Fighting on the question of reactor safety requires detailed technical knowledge – as the nuclear critics have found. Mounting a campaign against a drug, insecticide, or other environmental chemical thought to be dangerous, demands a thorough understanding of toxicology and other specialised areas of science. But just as the computer critics have developed techniques of guerrilla warfare to thwart the machines, so other critics of the applications of science and technology are using science and technology as weapons of attack. Nowadays, instead of simply complaining about noise, people use noise meters to chart noise profiles. Britain's Noise Abatement Society, for example, has encouraged such tactics in recent years. And when Concorde returned to Heathrow airport, London, in June 1972, after its world sales tour, there were environmentalists with noise meters there to meet it. Similarly, kits are now available to enable anyone to measure sulphur dioxide concentration in the atmosphere, and other forms of pollution. Armed with data, *as well as* arguments and tactical skill, the critic is in a much stronger position.

This trend is illustrated by the changing flavour of books on the environment published in recent years. After *Silent Spring*, a well-documented and reasoned book, there were many hysterical imitators. Purple prose replaced the facts – which are harrowing enough in themselves – as authors strained every sinew in an attempt to impress their readers. But recently the tone has altered again. Three books[21] published on behalf of the Friends of the

Earth in 1971 illustrate this clearly. Two of them, *The Environmental Handbook* and *The User's Guide to the Protection of the Environment*, contain detailed advice about practical action for individuals and communities, and include brand names etc. The third, the *SST and Sonic Boom Handbook*, is a well-documented source book explaining the known and potential hazards of supersonic aircraft, with practical advice for political action of the sort that helped to swell the political pressure that killed the American SST programme.

The military-industrial complex

Let us turn now to criticisms of science on a somewhat different plane – attacks on the military applications of science and the domination of science by industrial profit-making. They are linked via condemnation of the 'military-industrial complex', and indeed share many features in common – notably the intrusion of secrecy into science, which as we have seen is an intrinsically open subject, and the subordination of science to profit-making and warmongering. The degree of secrecy surrounding military research in so-called open societies is particularly staggering. For example, the first public mention of American interest in MIRVs (multiple independently targetable re-entry vehicles) was by Defense Secretary McNamara in an article in *Life* magazine[22] late in 1967. The development of MIRVs was then well in hand, and the final decision to deploy them was only a few months away.

One example of a specific campaign against military applications of science is the 'Honeywell Project', set up in Minneapolis (headquarters of Honeywell) in an attempt to halt the company's work on anti-personnel weapons and equipment for the automated battlefield (see p. 86). Vietnam war veterans, priests and students formed the original nucleus of the group when it was launched in 1968, but support grew quickly and in April 1970 3000 people demonstrated outside Honeywell's annual stockholders' meeting. The Honeywell Project has conducted extensive research into Honeywell's military contracts, publishing briefing documents describing techniques and equipment pioneered and produced by Honeywell. They also sponsored a 'Corporate War Crimes

Investigation' in February 1972 – the year in which offshoots of the Honeywell project were set up in Paris and London in attempts to make life uncomfortable for Honeywell associates and sub-contractors in other countries.

Similar to the Honeywell Project is the Polaroid Revolutionary Workers Movement, which in October 1970 called for an international boycott of all Polaroid instant cameras and gadgets, in protest against Polaroid's role as sole suppliers of instant pass photographs and race ID cards in South Africa. On Polaroid's own admission, the campaign was highly successful. By October 1971, the company's sales were down by 15 million dollars.

Rio Tinto-Zinc, with extensive mining interests in South Africa, has also found itself under heavy attack. When shareholders turned up to their annual general meeting in London in May 1972, they were handed copies of an 'anti-report', prepared by a group called Counter Information Services, which attacked RTZ not only for the 'dismal wages' paid to its workers in South Africa and its involvement in other parts of Africa, but also on environmental grounds – particularly the company's plans to mine in Britain's Snowdonia National Park. As with the Friends of the Earth books mentioned above, the 'anti-report' showed evidence of thorough research and careful preparation. 'The company conceals its true motivations, with their appalling social and environmental cost, behind a veneer of plausibility,' the report said. 'The purpose of this report is to tear away the mask and reveal the nature of the group through its actions and the way in which these affect us all.'

Another action was a campaign against 'Project Cambridge', based at the Massachusetts Institute of Technology. A 7·5 million dollar five-year programme funded almost entirely by the US Department of Defense, Project Cambridge was an attempt to use computerised behavioural modelling to help in the 'fight against Communism' in south east Asia. As Dr Joseph Hanlon explained in *New Scientist*[23] early in 1971: 'In the hills of North-East Thailand, anthropologists living in small villages gather data that will some day appear in scholarly papers. But the anthropologists *also* put their data into a computer at the Tribal Research Centre at Chiang Mai. The research centre has asked for information on village headmen, weapons on hand, and visitors to the villages. The data is available to anyone who is interested. The US military is interested, and is using the information to develop counter insurgency strategy.'

Behavioural scientists use mathematical models – sets of equations – to study groups of people in the same way that architects use models of buildings. Just as an architect's model shows what a building will look like when it is built, a behavioural model predicts how a group of people will behave at a later date. Moreover, as the architect can make structural changes to his design, so the behavioural scientist can use his model to determine how an outside influence is likely to alter the behaviour of his group. Project Cambridge included applications of this technique to such projects as studying 'mass unrest and political movements', analysis of interviews with the Vietcong, problems of development and stability in developing countries, and 'peasant attitudes and behaviour'. As applied to Thailand, the long-term aim was such that the military, using existing data, could predict which villages in a particular area were likely to be friendly or unfriendly, and what types of economic, social, military and political action would be most effective in gaining the support of the people. Using similar techniques, Hanlon argued, 'it would be possible in a few days to consider all possible strategies for an invasion of Cuba or North Vietnam. Probably further off, but still in the foreseeable future, such techniques could be used to decide whether or not to intervene in a foreign revolution or election.'

Not unexpectedly, this project – a typical example of the increasing military presence in the social sciences – attracted widespread criticism. Unimpressed by protagonists' claims that the use of computer modelling could have changed the course of the Vietnam war and avoided bloodshed, the critics argued that the technique was more likely to be used to compute the best way of achieving a military victory. Rather than curbing military action, it would facilitate it. Above all, critics pointed out the serious danger that reliance on computer modelling could lead planners into costly mistakes. And the critics had some success, impeding the project in a number of different ways. In 1969, for example, student demonstrations and public protest helped to persuade Harvard University not to become an official sponsor of the project.

To some critics, *all* so-called defence research is anathema. But a particularly common target is contract work done by universities and other bodies on behalf of the military. This too usually involves some degree of secrecy, but it is criticised largely on the grounds that it erodes independence and distorts university research (and thus teaching), making it dependent upon, and possibly subordinate to, the needs of the military. The sensitivity of this area can be gauged from the row which followed the disclosure in a British newspaper[24] in 1968 that several research groups in British universities were conducting work on contract for the Chemical Defence Experimental Establishment and Microbiological Research Establishment at Porton Down. Information about the contracts was given to the paper by a British MP who had visited Porton as a member of the Select Committee on Science and Technology (see p. 224). This was subsequently judged to be a breach of Parliamentary Privilege, and the MP was sacked from the committee and publicly reprimanded in the House of Commons. Although the Government later published the committee's report, some evidence was omitted on security grounds and apparently more was to have been left out but was included as a result of the newspaper leak.[25] Following the original report and the controversy it caused, student groups at several British universities named in the list of recipients of defence funds mounted protest meetings, and tried to persuade university authorities to decline contract work of this sort.

One of the most decisive recent campaigns against military research on campus was that at the Massachusetts Institute of Technology in 1969–70, which culminated in MIT's decision, announced on 20 May, 1970, to divest itself of its controversial Instrumentation Laboratory.[26] The brainchild of Charles Stark Draper – nicknamed 'Mr Gyro' after his work in developing the gyroscopic gunsight – the Laboratory became a target for intense student activity because of its involvement in such projects as MIRV development. A prime recipient of MIT's massive flow of defence research funds in the post-war years, the Instrumentation Laboratory had a budget of 54·6 million dollars in 1969, mostly from the Department of Defense and NASA. Trouble began in earnest with a 'research stoppage' on 4 March, 1969, organised by an uneasy alliance between the Science Action Co-ordinating Committee (a student group) and a group of staff and qualified scientists, the 'Union of Concerned Scientists'. Demonstrations, marches, petitions, and other action continued over the next year, and the MIT administration responded by appointing panels and committees of inquiry to investigate the disquiet surrounding the laboratory. Events reached a crescendo towards the end of 1969, and led to the final decision to separate the laboratory administratively from the Institute. While this was welcomed by several senior scientists at MIT who were disturbed by the laboratory's influence on university independence and integrity, it was students who had dragged the situation into public consciousness, forced many rank-and-file MIT scientists to consider their own allegiances in the matter, and precipitated the eventual outcome.

A related threat to the integrity of the university is posed by the growing collaboration between industry and university research workers. Again, this presents the possibility that, due to secrecy and the adjustment of research and teaching programmes to the requirements of outside commercial interests, the university as traditionally understood as a community of scholars and researchers will suffer. Such was the case with the row at the University of Warwick which came to a head in the winter of 1969–70.

In an article at the time,[27] E. P. Thompson, Reader in the History of Labour at Warwick, gave a hypothetical example of how corruption could develop in a university as a result of subordination to the needs of industry: 'A new university might be established . . . to which a new kind of vice-chancellor was appointed, who saw it as his particular mission to establish a new type of intimate (and subordinate) association with industry. He might see himself not so much as an academic organiser and arbitrator as the managing director of a business enterprise. Such a man would hold quite exceptional power and would have commanding influence over an input of some millions of money – mostly public, some private. Having a decisive say in the appointment of his first professors, he could also virtually nominate at his pleasure approved laymen to the university's governing council.

'If the university also raised several millions of pounds in private-appeal moneys, and if the allocation of these moneys lay effectively within his control, his ability to secure his own power by dispensing influence would be enlarged. It would be possible for him to decide – without regard to student demand, national need, or the logic of academic development and achievement – which areas should expand and which should be held back, what public image the university should project.'

Difficulties might arise with 'difficult' staff members who wanted a more open, democratic system, and with politically active students. But these could be overcome by such measures as using internal financial patronage to curb staff dissent and campaigns to blacken certain staff and students as 'disloyal' to the organisation. 'Such a corruption of purposes of a university could take place . . . without any violent fracture of legal proprieties and even without irregularities which call for the attention of auditors.'

This is a hypothetical case, but elements of the story are recognisable in real life cases that have come to light in recent years. At Warwick itself, for example, after five years of development the arts and several social science subjects were still housed in unsatisfactory, crowded accommodation, and the students' living, eating, and recreational facilities left much to be desired. But the sciences – thanks to money from industry – had substantial, well-equipped buildings. With projects on such problems as metal fatigue (financed by Massey Ferguson), vehicle instrumentation (Rootes and Ford Motor Company), fatigue in tyres (Dunlop), and high speed machine-tool cutting tips (Alfred Herbert), industry looked upon the university largely as a laboratory for its own research and development work. As *The Times* technology correspondent wrote at the time, Warwick was 'tailored to industry's needs'. Of the nine co-opted members of the university council, eight were representatives of substantial business interests. (The ninth was a Bishop.)

Unrest about the way Warwick was run came to a head on 11 February, 1970, when students, fed up after three years of trying to gain control of their own student union (student control was commonplace at other British universities), invaded the registry. There they found an astonishing report by the Director of Legal Affairs of the Rootes Organisation on a visiting American professor, David Montgomery. He had, apparently, been asked to report to the vice-chancellor on the professor's political activities. In other words, a university official had used a legal expert employed by a large industrial corporation to advise on an academic colleague whose political complexion gave rise to anxiety. 'This is, of course, the corporate society, with all its ways of adapting and tailoring men to industry's needs, the corporate managerial society, with its direct access to legal process to prevent the truth from being published,' wrote Thompson, commenting on this and other incidents at Warwick. 'It might be thought that we have here already, very nearly, the private university, in symbiotic relationship with the aims and ethos of industrial capitalism, but built within the shell of public money and public legitimation. . . . We have been luckier than any of us had the right to deserve in the quality of our students. They asked the right questions. They began to understand the answers.'

No-one complains about greater intercourse between universities and the outside world; all too often the work and teaching of a university is out of gear with the needs of society. But in the case of contract research work for science-based industry, the intercourse is very much with just one sector – industry, capitalism, profits, management. What the critics would like to see is greater relevance of research pursued in the university to the social needs of the people (as reflected in the name of the American publication *Science for the People* mentioned on p. 184). The same applies to science education. As we have seen (p. 55), typical science degree courses simply ignore the social, political, and economic effects of science. They even neglect the real social and economic background to the historical development of science itself. Again, students have been in the forefront of criticism, calling for greater relevance of science curricula to the real world, less emphasis on rote learning of facts, and a more open-ended approach to problems to replace the digestion of slabs of information.

Do we want everlasting growth?

At this point, we encounter criticism of science on a considerably wider basis than the specific examples we have discussed in this chapter. The demand for greater relevance in science teaching leads to questions about the type of society in which we live and the role of science and technology in that society. Hence to those critics such as Dr Edward Mishan who question the desirability – even possibility – of everlasting economic growth, with its dependence on continuous industrialisation and technological development.[28] In the nineteenth century, Mishan argues, sustained growth and free trade were the dominant aspirations, but changes arising from the gathering pace of science and technology have thrown the whole doctrine of continuous growth into doubt. The problem arises largely from the unprecedented population explosion (with its damaging ecological consequences); the post-war surge of affluence in the west, largely channelled into communications, the growth of car ownership, travel, etc.; and the speed of technological advance, which creates problems of obsolescence of goods and skills. 'The pursuit of efficiency', he argues, 'itself regarded as the lifeblood of progress, is directed towards replacing the dependence of people on each other and increasing their dependence on the machine. Indeed, by gradually displacing human effort from every aspect of living, technology will eventually enable us to slip swiftly through our allotted years with scarcely enough sense of physical friction to be certain we are still alive.'[29]

Alvin Toffler, in his book *Future Shock*,[30] also writes of 'the roaring current of change, a current so powerful today that it overturns institutions, shifts our values and shrivels our roots'. We live at a time of not only technological and scientific revolution but also youth revolution, sexual revolution, radical revolution and colonial revolution, yet we scarcely have time to understand the significance of these events or where they are taking us. There is 'sheer factual ignorance about the nature, meaning and direction of the super-industrial revolution'.

In their different ways, Mishan and Toffler both point to two directions of further inquiry into the profound changes now taking place in the world as a direct result of the 'success' of science and technology. The first, paradoxically, calls for more science, not less. Caught in a crescendo of change, we desperately need to *quantify* the various trends of population, pollution, and economic change, and to attempt to predict the future pattern. Neither pontifications of doom nor bland reassurances are acceptable answers in this situation. Hence the importance of programmes such as the MIT computer-modelling study to define the 'limits of growth' (see p. 208) and similar exercises designed to replace hysteria by facts and soundly-based theory.

The second need is for thoughtful analysis of the ways in which our profound dependence on – indeed domination by – science and technology has changed the quality of human relationships and the texture of society. Toffler is right. We are busy. We do not stop to reflect any more. We are seduced by the process of change itself. That is why some of the most radical and disturbing criticisms of science and technology come from those who simply tell us, carefully and coolly, what they have done to us. The most discerning of these contemporary critics of science and the scientific world view is the sociologist Theodore Roszak, and it is to his critique that we now turn.

Chapter 9
Reason in perspective

Stinking rivers, filth in the air we breathe, omnipresent noise, the plunder of raw materials, weapons of devilish savagery – all these bear witness to the dark face of science and technology. Despite attempts by the experts to persuade us that such horrors are merely temporary problems thrown up in the course of progress, people have recently begun to rebel. The products and processes of science and technology are under sustained attack. Yet, seen on a broader canvas, there are even more serious allegations against science on a different level altogether. The crucial criticism – all the more potent because we are seldom consciously aware of the case that supports it – is of the extent to which science dominates our lives, our 'world-view', habits of thought, human relationships, and values – our entire cradle-to-grave existence. Science-based industrialism has virtually obliterated any alternative style of life, and we have scarcely even noticed.

As with that brainteaser about a three-dimensional man trying to explain his world to people living in only two dimensions, our own domination by science makes it difficult to discern the substance behind these criticisms. The break-neck rate of technological change renders matters even worse; we simply have no time for the luxury of philosophical contemplation. By comparison, taking up cudgels against a proposed nuclear power station, or fighting a political battle over the amount of carbon monoxide emitted by automobiles, are much easier propositions. Such games are played according to the well-understood rules of the technocratic society – as we have seen, ammunition in the form of technical arguments is more and more necessary on both sides. Indeed, the technocrats will probably increasingly welcome well-informed, technical criticism, as a stimulus to more ingenious innovation, and will settle down to a relatively comfortable existence with the more professional activist groups. If not exactly cosy, the relationship will be one of mutual respect and comparable professionalism, like employers and trade unions in a well-adjusted mid-industrial society. (Ralph Nader, for example, is by no means as unpopular with American business as many accounts of his activities would have one believe.)

By contrast, there is virtually no communication whatever between professional scientists and technologists on the one hand and on the other people such as Jacques Ellul, Herbert Marcuse, and Theodore Roszak, who confront science at a more comprehensive and profoundly disturbing level. Any debate that does take place usually ends in frustrated incomprehension, with the professionals simply not understanding the critics' criticisms. This is not surprising; the professionals are soaked in the assumptions of science (as are we all to a degree) but have the additional disadvantage of a narrow technical education and membership of a community dedicated to reductionism (see p. 178) and conformity. It is a particularly dangerous situation, the more so because people like Roszak are followed by countless young disciples who are seduced by their erudition and the potency of their attack on the life-styles of today, but who lack their wisdom. 'Anti-science' fostered by movements of this sort can easily degenerate into nihilism, mere silliness, or a slavish devotion to intuition at the expense of the intellect – which has pernicious political undertones.

Although such movements have burgeoned in recent years, the ideas behind them are not new. Before turning to the contemporary prophets of anti-science, therefore, it will be useful to examine their historical antecedents. Stephen Toulmin, speaking at the Ciba Foundation symposium mentioned in the previous chapter, identified five themes which recur again and again at every stage of the anti-science debate.[1] These are humanism, individualism, imagination, 'quality versus quantity', and the abstract character of scientific ideas and inquiries.

The idea of humanism can be traced back to Socrates, who emphasised the primacy of social, ethical and humane issues, and was healthily sceptical about the possibility of totally reliable, objective scientific knowledge. So it was with Michel de Montaigne and the humanists of the 16th century. While the scientific ideas of the classical Greeks had survived through the Middle Ages, the humanists had to rediscover the poets, essayists, historians, and tragedians of antiquity. In doing so, they made a unique contribution to European sensibility. 'The humanists tended – like the romantics of the early 19th century – to go on and pillory the scientists for being indifferent, and even callous, about humane issues,' writes Toulmin. They did so 'with the same kind of passion as any of today's anti-scientists denouncing nerve-gas research or the alliance between official science and the military-industrial complex.'

Individualism is another characteristic claimed uniquely for literature and the arts by the enemies of science, because these activities give scope to the individual personality of the writer or artist. In contrast, science is looked upon as a conformist activity in which research workers rigidly suppress their personal and subjective views and feelings in favour of communally imposed orthodoxy. As we have seen, this is by no means totally true, but the comparison with the arts is accurate and scientists themselves certainly make a virtue of the objective character of their work. Closely connected with the theme of individualism is that of imagination. Science (the argument runs) is based on mechanical and stereotyped modes of inquiry, which tend to starve creativity and the imagination. Again, although much normal, bread-and-butter science is pursued on this basis, creativity akin to that of the artist *is* important at the major turning-points of research, but scientists disguise the fact when presenting and discussing their research findings. To that extent, scientists are indeed peculiar people, who make a virtue of denying their own creativity.

The fourth theme, quality versus quantity, was emphasised by Goethe in particular. It stems from scientific method. The scientist concerns himself only with the common features shared by many individual things, organisms, phenomena – and in the last resort with statistical averages and measurable units. This leads him to neglect individual differences and to ignore qualitative variations. Perhaps the starkest modern example is the sociology of mass observation, according to which one can gain an exhaustive account of society and groups within society simply by counting heads, totting up salaries, sampling opinions by national opinion poll. (We encountered the same technique as applied to scientists themselves on page 44.) One of the greatest contributions to this interpretation of science was Isaac Newton's *Principia Mathematica*

Philosophicae Naturalis, a quarter of a million words on the laws of motion, the mathematical analysis of motion, and the movement of heavenly bodies, which he believed to have established that the universe was made by a Rational Being. As seen by the romantics, writes Toulmin, Newtonian science was guilty of 'ignoring the individual, of sub-ordinating qualitative differences to quantitative uniformities, of killing the animal whose life it pretended to explain, of breaking up into a spectrum (and so destroying) the whiteness of light, which needed rather to be studied in its primal integrity, and so on.' On the romantic view, scientists should cultivate more of the personal, humane insight into the world which good doctors bring to their individual patients. They should become not objective automata but artists of the intellect, 'developing a feeling for the personality and uniqueness of each colour, leaf, human being, or meteorological event, without which he could not seriously claim to "understand it" '.

The abstract character of scientific investigation is a related complaint. It is the one which probably comes closest to identifying the psychological traits which drive individuals to become scientists (if, as I suspect, such traits do exist). Scientists in general are unhappy about the untidiness of political, social and personal relationships in the real world, and find it difficult to take the actual course of events, or a particular problem, as they find it. 'They begin by imposing certain arbitrary theoretical demands and standards on the variety of nature, and they are then prepared to pay serious attention only to those aspects of nature which they choose to accept as "significant" by those standards,' argues Toulmin. 'A true humanism, by contrast, will be prepared to accept each new concrete situation in all its complexity and variety, as it arises, and deal with it accordingly.' It is this deficiency which has led to so many attacks on scientists and technologists for callous indifference to the broad, human implications of their work. For a group of scientists, it can be self-evident that their work is beneficial and should be accepted as such by the people likely to be affected. This is true almost irrespective of the actual merits and demerits of the work in question. Poring over their maps and rainfall figures, it can be self-evident to the senior scientists of a chemical company that their new plant, bringing employment to an area by producing millions of plastic milk bottles, must have more water; that therefore a new reservoir will have to be constructed in a particular valley; and that the residents there will simply have to move. Scientists actually trying to be directly helpful can be equally blind to human factors – in offering a new synthetic food to the people of a developing country, for example, or proposing a daring new surgical operation to the anxious parents of a seriously ill child.

Each of the five different limbs of the anti-science movement can, then, be traced back into history. With the tremendous successes of science over the past hundred years, however, they have been increasingly overshadowed. In particular, the revolutions wrought by science in medicine, agriculture, and manufacturing industry have persuaded people that science is a 'good thing'. More than that, because science has proved itself so overwhelmingly, we tend nowadays to elevate the needs and explanations of science, quite wrongly, to a very high level indeed. As Michael Polanyi says: 'In the days when an idea could be silenced by showing that it was contrary to religion, theology was the greatest single source of fallacies. Today, when any human thought can be discredited by branding it as unscientific, the power exercised previously by theology has passed over to science; hence science has become in its turn the greatest single source of error.'[2]

Suddenly, in the past ten years or so, we have woken up to this state of affairs and feel angry about it. That, alongside growing awareness of environmental damage attributed to science, is the origin of the current attacks on the scientific establishment. We have found ourselves dominated by what Theodore Roszak, in *The Making of a Counter Culture*,[3] calls 'the technocracy . . . that social form in which an industrial society reaches the peak of its organisational integration. It is the ideal men usually have in mind when they speak of modernising, up-dating, rationalising, planning. Drawing upon such unquestionable imperatives as the demand for efficiency, for social security, for large-scale co-ordination of men and resources, for ever higher levels of affluence and ever more impressive manifestations of collective human power, the technocracy works to knit together the anachronistic gaps and fissures of the industrial society. The meticulous systematisation Adam Smith once celebrated in his well-known pin factory now extends to all areas of life, giving us human organisation that matches the precision of our mechanistic organisation. So we arrive at the era of social engineering in which entrepreneurial talent broadens its province to orchestrate the total human context which surrounds the industrial complex.'

Here Roszak joins forces with another critic of contemporary science-based civilisation, Jacques Ellul, in attacking our preoccupation with *technique*. For Ellul, technique is not just machine technology, but covers also the standardisation of procedures and behaviour to develop 'the one best method' for achieving a particular result, to the detriment of spontaneity. In place of unreflective, spontaneous action, it imposes deliberate rationality. Just as Robert Oppenheimer, one of the leading physicists in the American controversy over whether or not to build the H-bomb, described his work at Los Alamos as 'technically sweet',[4] so the Technical Man cannot help but admire the spectacular effectiveness of weapons of war. 'No technique is possible when men are free,' writes Ellul.[5] 'When technique enters into the realm of social life, it collides ceaselessly with the human being to the degree that the combination of man and technique is unavoidable, and that technical action necessarily results in a determined result. Technique requires predictability and, no less, exactness of prediction. It is necessary, then, that technique prevails over the human being. For technique, this is a matter of life or death. Technique must reduce man to a technical animal, the king of the slaves of technique. Human caprice crumbles before this necessity; there can be no human autonomy in the face of technical autonomy. The individual must be fashioned by techniques, either negatively (by the techniques of understanding man) or positively (by the adaptation of man to the technical framework), in order to wipe out the blots his personal determination introduces into the perfect design of the organisation.'

Both Roszak and Ellul perceive that no area of life is free from the pressures and demands of the technocracy. Everything becomes the preserve of the technical expert – sexual behaviour, child-rearing, and recreation being some of the currently popular areas receiving enormous technical scrutiny. Abraham Maslow[6] gives one delightful example which highlights the schism between the blundering, systematising expert and 'ordinary' people. He quotes a scientist who praised a book on 'the difficult problem of woman's sexuality', because at last it took up a subject 'about which so little is known'.* In its most highly developed form, technique appears as the multi-billion-dollar 'think-tank' in which cadres of scientific and technical experts study, scrutinise and forecast every real and conceivable facet of society, in an attempt to anticipate and integrate absolutely *everything* that exists or might exist in society, in the name of rational, purposeful planning. 'Within such a society, the citizen, confronted by bewildering bigness and complexity, finds it necessary to defer on all matters to those who know better,' writes Roszak. 'Indeed, it would be a violation of reason to do otherwise, since it is universally agreed that the prime goal of the society is to keep the productive apparatus turning over efficiently. In the absence of expertise, the great mechanism would surely bog down, leaving us in the midst of chaos and poverty.' In the long term, 'the capacity of our emerging technocratic paradise to denature the imagination by appropriating to itself the whole meaning of Reason, Reality, Progress, and Knowledge will render it impossible for men to give any name to their bothersomely unfulfilled potentialities but that of madness. And for such madness, humanitarian therapies will be generously provided.'

Behind these criticisms lies what Roszak calls 'the myth of objective consciousness. There is but one way of gaining access to reality – so the myth holds – and this is to cultivate a state of consciousness cleansed of all subjective distortion, all personal involvement. What flows from this state of consciousness qualifies as knowledge and nothing else does. This is the bedrock on which the natural sciences have built; and under their spell all fields of knowledge strive to become scientific. The study of man in his social, economic, psychological, historical aspects – all this, too, must become objective: rigorously, painstakingly objective.'

The fact that such critiques of 'objectivity' sound like so much gibberish to many scientists reflects the philosophically barren nature of scientific training. Very, very few scientists appear to be aware that *all* experience – whether of meter readings in a laboratory or the love of one person for another – is subjective. There are

* This reminds me irresistibly of James Cameron, writing in *Punch* (2 April, 1972) about scientists: 'Should they be trusted at all? . . . With one's wife, probably yes, with one's life emphatically not.'

qualifications and assumptions underlying all the impregnably certain data of science, with no convincing reason whatever why so-called exact knowledge obtained by 'the scientific method' should always claim our prior assent over other sorts of experience. The logical, objective, analytical 'I' controls – can control – only a part of my existence, and even the data with which I might try to discredit the more obviously subjective experiences and beliefs are acquired by courtesy of sense impressions. Michael Polanyi stresses this neglected fact about our means of knowing in this passage from his *Personal Knowledge*[7]:

'Objectivism has totally falsified our conception of truth, by exalting what we can know and prove, while covering up with ambiguous utterances all that we know and *cannot* prove, even though the latter knowledge underlies, and must ultimately set its seal to, all that we *can* prove. In trying to restrict our minds to the few things that are demonstrable, and therefore explicitly dubitable, it has overlooked the a-critical choices which determine the whole being of our minds and has rendered us incapable of acknowledging these vital choices.'

In short, there is no reason why one should dismiss experience that does not accord with one's theoretical constructs about 'the outside world', nor reject intuitive feelings out of court because they are illogical or inexact. In building a philosophy, one has no right to dismiss some categories of experience because others have overall priority. The task is to evaluate them together and combine them into a satisfying whole. (We should never forget, either, that no series of 'third person' accounts of a phenomenon can ever replace a 'first person' account.)[8] It is difficult to over-emphasise the importance of placing 'objectivity' firmly in its place.[9] Scientists themselves tend to be ignorant of the philosophical pitfalls in this area, and it is not therefore surprising that – with their considerable practical achievements and confident pronouncements – they should have helped to keep the rest of us in the dark too.

At this point, it is important to answer the reasonable charge that the type of argument outlined above implies licence to suspend critical judgment and believe any old nonsense. In particular, it can be argued that such a standpoint leads naturally to anti-intellectualism, romanticism, mysticism, and ultimately political

Fascism. J. D. Bernal, in *The Social Function of Science*,[10] claimed that 'metaphysical scientists' and philosophers such as Henri Bergson and Georges Sorel helped to pave the way for the justification of the Fascist ideology of brute force under mystically inspired leadership. More recently Sir Eric Ashby has raised the same spectre in attacking those who 'question the very legitimacy of scientific thought'.[11] Clearly, any philosophical position – from subjective idealism to existentialism or materialism – driven to its extreme limits and accepted with single-minded passion as the one and only guide to life and political action, could have dangerous consequences. In this century alone, we have seen wars and movements of oppression drawing their strength from a wide range of different philosophies and ideologies. But a crude materialism is as likely to lead to callous indifference, as is mysticism to predicate Fascism. Many of those who criticise people such as Roszak on this basis are missing the point. Theodore Roszak does not prescribe mysticism to the neglect of science. What he does argue is that we should place the scientific world view – and objective consciousness in particular – in its balanced perspective, and move away from the grossly exaggerated reverence for science which is now current. For Roszak, objective consciousness is 'an arbitrary construct in which a given society in a given historical situation has invested its sense of meaningfulness and value.'

So it is that, in today's counter-culture, we find people rejecting science, scientists, and all they stand for. They do so because they have called science into question and found meaning and value elsewhere. Hence the great interest aroused by such books as *The Teachings of Don Juan* and *A Separate Reality*,[12] in which Carlos Castaneda records his conversations with a Yaqui *brujo* (medicine man) who taught him to explore 'non-ordinary' reality. Hence the burgeoning following for unfamiliar, non-institutional forms of religion. Not the opium of the people, but the 'poetry of the people',[13] religion is welcomed as wholesome, poetic refreshment to compensate for the barren, cheerless world moulded by science. Hence the 10,000 professional astrologers working today in the United States, compared with a mere 2000 astronomers.[14] Hence the manifesto pinned to the main entrance of the Sorbonne during student troubles in May 1968: 'The revolution which is beginning will call in question not only the capitalist society but industrial society. . . . We are inventing a new and original world. Imagination is seizing power.'[15] Hence the contemporary popularity of psychedelic drugs, even among apparently 'well adjusted and integrated' middle-class American adults. Hence the much sought-after gurus, yogis, prophets and oracles dispensing the wisdom of the East. Hence the apparently negative tone of so much of the current movements of counter-culture. As Lewis Mumford observes: 'Since ritual order has now largely passed into mechanical order, the present revolt of the younger generation against the machine has made a practice of promoting disorder and randomness . . .'[16]

There is one great problem. The message of Theodore Roszak and those who argue along similar lines would be even more potent if they were not so totally opposed to science in its every manifestation. Roszak diminishes science. He doesn't seem to understand its truly creative and aesthetic aspects or its positive value as a cultural force. Nor – even though he mentions the possibility in the opening pages of *The Making of a Counter Culture* – has he much time, apparently, for the capacity of the products of the technocracy to liberate the human body and spirit. He disparages molecular biology ('Consider the strange compulsion our biologists have to synthesise life in a test tube – and the seriousness with which the project is taken. Every dumb beast of the earth knows without thinking once about it how to create life'), but ignores its potential capacity to banish the misery and senseless losses wrought by cancer or to help a woman with a serious hereditary disease to bear a normal child. He forgets that, despite the personal greed it exploits and the material waste, pollution, human carnage, and urban and social problems it creates, the motor car can also help to promote personal relationships and fulfilment. As mentioned in Chapter 5, even television, for all its dross and pabulum, has the immensely valuable capacity of widening our range of experience and our knowledge of the world in which we live.

In short, Roszak's uncomfortably accurate diagnosis of our thraldom to the technocracy is bound to appear, to many of those whom he most wants to influence, as an exaggerated and short-sighted attack from a position of privileged security. He has, indeed, added one more brick to those very fortifications of entrenched complacency he attacks with such dedication. True, prophets have always overstated their case – but they have never had to deal with the technocracy before.

Much the same applies to the attacks on our technological age voiced by a very different critic, Dr F. R. Leavis, the Cambridge don and literary critic. A self-elected custodian of austere standards of academic scholarship, he has had a deep influence on the literary scene in Britain; and it is a sobering thought that so much of his case against our domination by science should coincide with that promulgated by the young spokesmen of the counter-culture. 'The educated and cultivated have, in general, surrendered to the climate of the technological age,' Leavis claims.[17] As a result, life is being impoverished while education, harnessed to material growth, is being sullied by crude considerations such as cost-efficiency. Standards – *real* standards – are in jeopardy because 'the drive of our technological-Benthamite world is not merely indifferent, but hostile, to the human creativity they represent'.

It is acceptable, Leavis concedes, to subject society's problems to scientific scrutiny 'and deal with them in terms of statistical data, charts, and the computerisable generality; we are committed irrevocably to the necessity of government, government departments, complex machinery of administration, and bureaucrats.' But 'the "social" has to be conceived in another way' and we must seek to preserve 'a conception of society, life and *humanitas* that doesn't eliminate the depth in time and the organic'.

Again, Leavis is profoundly right in his desire that life should not be impoverished by the outpourings of science. But one of the reasons why he writes in this shrill fashion is because he has not understood the essence of science, its creativity, its cultural importance, or the nature of its interactions with society. Platitudinous the observation may be, but we must remember that the computer can be a master or slave (Leavis was particularly incensed by a colleague who told him that a computer could write a poem); that using one pair of analytical spectacles does not preclude changing to another pair for a different purpose; that modern technology, though it may have attenuated certain harmonies in the texture of society, has also heightened people's awareness of the world and increased their capacity to change it; and that despite the loss of certain simple pleasures, and the irritations of a phrenetic existence, life today in the Western world is healthier, and society more humane, than was the case even 100 years ago. These are not insignificant achievements, and to neglect or minimise them only weakens the case against the positive losses we have incurred as a result of scientific progress.

Returning to the philosophical basis of the anti-science lobby, a unique recent example of the scientific world view in all its disturbing clarity is to be found in *Chance and Necessity* by Professor Jacques Monod.[18] A *tour de force* in which a distinguished biologist deploys the anti-religious world-view of modern biology in all its weight and detail, Monod's book has attracted widespread criticism and even ridicule from the anti-science camps. Jacques Monod, together with François Jacob and André Lwoff, received a Nobel Prize in 1965 for charting the control mechanisms which regulate living processes. Even in such humble creatures as the bacteria living in the human bowel, these controls are complex and exquisitely sensitive. One of their purposes is to link together and regulate the chemical reactions which break down food materials in cells and those by which the bacterium builds up new cell materials. Without such control mechanisms, growth would be unbalanced and the cells could not adapt to changing external conditions.

Part of Monod's case in *Chance and Necessity* rests on such 'telenomic' mechanisms – those in which apparently purposive, goal-directed behaviour suggests design and purpose but which can be adequately explained in terms of physics and chemistry. Despite persuasive evidence to the contrary, Monod argues, the entire phenomenon of life on earth, from its origin to the evolution of man, is based on random, chance happenings in chemical molecules. The process began with a fortuitous concourse of atoms in the primeval soup and has continued by the blind forces of Darwinian natural selection acting on mutations thrown up by equally blind, random forces. There is 'pure chance, absolutely free but blind, at the very root of the stupendous edifice of evolution'. As a result of the past hundred years of biological research 'man now knows at last that he is alone in the universe's unfeeling immensity, out of which he emerged only by chance'.

The historical shift from Darwin's championing of natural selection to Monod's preoccupation with randomness in nature is one of those cases where, on classic Marxist lines, quantitative change has led to qualitative change. Darwin emphasised the aimlessness of natural selection, but his theory of evolution left room for a belief in design and creation. Monod spells out in considerable detail the inescapable influence of chance on the fabric of life, at many different levels of biological organisation. The evidence at his command allows him to deal cruel death blows to such concepts as vitalism and panpsychism (according to which the antecedents of mind and consciousness are to be found in every crumb of matter in the universe), and indeed to make any assertion of religious belief appear totally absurd.

Not only that. If Monod is right, all of the ideological products of Western civilisation – including our concepts and institutions of justice and education – go into the melting pot. 'For their moral bases,' he argues, 'the "liberal" societies of the West still teach – or pay lip-service to – a disgusting farrago of Judeo-Christian religiosity, scientistic progressism, belief in the "natural" rights of man, and utilitarian pragmatism.' The *Weltanschauung* of Jacques Monod – which is also the world view accepted implicitly by the vast majority of all working scientists today – provides no support whatever for the disgusting farrago, and every reason to jettison the entire fabrication. However, Monod argues, if people do endorse 'the essential message of science', then man can at last burst free from his millenary dream. 'He wakes at last to the realisation that, like a gypsy, he lives on the margin of an alien world. A world that is deaf to his music, just as indifferent to his hopes as it is to his suffering or his crimes.'

Needless to say, this is all anathema to Roszak, who writes:[19] 'How sad to be Jacques Monod: to stare so expertly at miracles and meanings, and never see them.' Yet even Roszak seems so overwhelmed by Monod's unrelenting exercise in philosophical tough-mindedness that he fails on this occasion to call the scientist's bluff. There are, in fact, two fundamental snags about Monod's bleak diagnosis. One we considered above – the fallacy of attaching cardinal importance to factual, scientific knowledge as against other experience. The second point is that the scientific world-view is by no means as free of anomalies and paradoxes as Monod presents it, chance notwithstanding, in his book. True, there is apparent randomness in the behaviour of the molecules which mediate living processes. But when we scrutinise matter intimately, in an attempt to understand its own nature, when we look in particular at the sub-atomic world, we encounter phenomena which are antithetical to common sense. The simultaneous behaviour of electrons as both waves and particles is only one of several totally perverse paradoxes which scientists are prepared – nay compelled – to accept as a consequence of their own research. The field of relativity provides several more bizarre phenomena which, in common sense terms, amount to total paradox. Why, then, should not a religious believer cheerfully accept the paradox posed by the apparent conflict between his own faith and Monod's capricious materialism?

The last few years have in fact seen the first stirrings of discontent within the scientific community itself over the domination of biology by reductionism, exemplified by Monod's position. In 1969, for example, the important book *Beyond Reductionism*[20] appeared in which fifteen distinguished scientists from a range of different disciplines discussed the inadequacy of reductionist explanations in their various fields. As summarised by the animal ethologist Professor W. H. Thorpe,[21] the targets of their attacks were four 'pillars of unwisdom': 1. that biological evolution is the result of nothing but random mutations preserved by natural selection; 2. that mental evolution is the result of nothing but 'random tries' preserved by 'reinforcements' (as occurs when a rat is rewarded with food when it presses a particular bar in a cage and is punished by an electrical shock when it pushes the wrong lever); 3. that all organisms, including Man, are nothing but passive automata controlled by the environment, whose sole purpose in life is to reduce tensions by adaptive responses;* and 4. that the only scientific method worth that name is quantitative measurement; and, consequently, that complex phenomena must be reduced to simple elements accessible to such treatment – without undue worry whether the specific characteristics of a complex phenomenon – Man, for instance – may be lost in the process.

* B. F. Skinner, in *Beyond Freedom and Dignity* (Knopf, 1971) argues not only that free will is an illusion, but that we must control man more efficiently in future, designing a culture in which man will refrain from polluting, over-populating, and fighting because he has been conditioned to serve group interests.

In attacking this creed, Professor Thorpe and others do not in any way allege that reductionism as a *technique* is either reprehensible or avoidable. As we have seen, to further scientific understanding, one must reduce problems to their simplest essentials. The error is in narrowing one's vision and treating an explanation afforded by such atomistic dissection as an exhaustive explanation – as, for example, with those psychologists who see nothing in human behaviour and consciousness not explainable on the basis of their behavioural studies with rats, and thus concoct what Arthur Koestler calls a 'ratomorphic' view of man. More and more, biologists find that they must study systems, seek pattern in whole organisms, and look for synthesis rather than analysis, if they are truly to understand the phenomena they are dealing with. This has enormous implications for the larger philosophical questions of human values. 'If reductionism were right in the sense that the mental, spiritual, and ethical values which we experience really *are* in the electrons and other primary components of which the world is made,' writes Thorpe,[21] 'then all one can say is that they don't *appear* to be there. It follows that a great and unjustified leap of faith is required, a leap without any scientific evidence, to believe it.'

It is possible, then, that slowly and against the great inertia of the conventional wisdom, the domination of biology by reductionism may be purged from within simply by dire necessity. Even as a tool, reductionism sometimes fails, and science must seek broader patterns of meaning. There is a parallel here with the measure of 'reform from within' which has helped to change attitudes towards the environment in recent years. Scientists have learned the hard way that it is foolhardy to neglect ecology in tampering with the environment. Alter just one tiny part of a natural ecosystem built up over hundreds of thousands of years, and you risk dangerous repercussions, perhaps in an area remote from the one with which you interfere. Quite apart from the lamentations and warnings of the environmentalists, who believe on much broader grounds that the environment should not be impaired, there is substantial self-interest in treading carefully and considering the whole before disturbing the part.

But this comes nowhere near to a satisfactory solution of the environmental, philosophical or social problems arising from science and technology. Selfishness is scarcely the most lofty of motives, and there is a danger that, far from awakening wider concerns and sensitivity, the necessity to take account of broader parameters than those now considered orthodox may simply exaggerate a pragmatism which says: that which is true is that which works and is useful.

Chapter 10
Dissent and disquiet from within

One afternoon in December 1971 in Philadelphia, participants at a session of the annual meeting of the American Association for the Advancement of Science (AAAS) were listening to a lecture on 'The Immunity and Immunopathology of Oral Soft Tissues'. The light was out and they were peering at slides showing the microscopic features of dental decay. Suddenly, the door broke open and fifteen or so members of an organisation called Scientists and Engineers for Social and Political Action (SESPA) stumbled in and began handing out leaflets about a forthcoming vigil and march of protest against the Vietnam war. There were shouts of 'Down in front' and 'Get them out of here', and the invaders were physically removed. Muttering to each other about the incongruous interruption, the dentists returned to their pathology slides.

Elsewhere at the giant AAAS meeting, other 'flying squads' from SESPA were more successful in persuading scientists to join their protests, organised by Vietnam Veterans against the War. Attempts to introduce other political issues into the meeting also succeeded, often against initial opposition from participants and chairman at the scientific sessions. During a symposium on 'Technology and the Humanisation of Work', for example, an official from a telephone company used a chart with cartoon figures to illustrate his talk on customer service. The chart showed managers as male, and the lower grades of supervisor and service representative as female. 'I want to know just exactly what percentage of women are employed as workers and what percentage as managers,' demanded a SESPA representative in the audience. Pandemonium broke out. Someone tried to put his hand over the mouth of the person who asked the question. The chairman shouted something about disruption and the chance to ask questions later. The speaker tried to speak. Then, when things began to quieten down, first one and then a number of non-SESPA scientists in the audience asked to have the question answered. The chairman agreed, and the speaker conceded that although some 90 per cent of workers in his company were women, there were virtually no female managers, and that this did suggest a degree of sex discrimination. A worthwhile exchange of views followed, and just before the session adjourned several of the panellists thanked the SESPA members for coming to the meeting, remarking that, contrary to first impressions, their participation had been both sincere and constructive.[1]

SESPA is just one of a number of groups concerned with the social relations of science which have grown up in recent years within the scientific community. They differ widely in ideology and methods, but are united in urging that scientists should be politically and socially active, seeing their science as an integral part of other human affairs. The activist groups tend to concentrate on such issues as defence, environmental pollution, and economic structure, plus particular, politically explosive topics such as the relationship between intelligence and race. But the main plank of their case is that no segment of science is neutral. Their crusade – for that is what it amounts to – is aimed at awakening the mass of working scientists to the social and political implications of science, making it impossible, or at least uncomfortable, for a scientist to ignore the wider context of his work and his obligations towards the society which supports him. As formulated by two of the British leaders of the movement for social responsibility in science, scientists should ask themselves the question: 'How can my scientific skills be best used to serve the people: to expose and correct the role of science and technology in wreaking genocide in wars, in oppressing individuals and minorities by acting as an agency of civil war, and in permitting malnutrition and disease both throughout the world at large, and even in rich societies?'[2]

Despite the belief of some present day activists that they have discovered social responsibility in science for the first time, the idea has important historical roots. The distinguished pathologist Rudolph Virchow, for example, led a movement among German scientists in the mid nineteenth century designed to democratise and humanise science. An attempt to stir the consciences of researchers, it seems to have started as a reaction against the complaints of the romantics about science earlier in the century.

Thereafter, numerous small groups appeared and later waned – typified by the Cambridge Scientists' Anti-war Group, set up in the early 1930s to study the causes of war and the measures scientists specifically could take to help preserve the peace, and the similar *Comité de Vigilance* in France. In 1939, Bernal's *The Social Function of Science* was a major attempt to stir the social conscience of scientists everywhere. But it was the involvement of scientists in the Manhattan project to produce the atomic bomb, and their reactions afterwards – 'The physicists have known sin, and this is a knowledge which they cannot lose,' said Robert Oppenheimer – which gave the greatest spur to movements designed to awaken social responsibility among scientists. After the war, Professor Eugene Rabinovitch, Dr Leo Szilard and others founded the Federation of Atomic Scientists (now called the Federation of American Scientists) which, through its publication the *Bulletin of the Atomic Scientists*, tried to prevent the spread of nuclear weapons and to place the development of nuclear power outside the charge of the military. The McMahon Act, which secured nuclear weapons development under the non-military US Atomic Energy Commission, was the first real success of this lobby.

Later, with growing concern during the 1950s over atomic weapons testing, came the beginning of the Pugwash organisation,[3] which held its first meeting at Pugwash, Nova Scotia, in July 1957. With financial help from the industrialist Cyrus Eaton, its inception followed the famous manifesto drafted by Bertrand Russell and signed by Albert Einstein warning of the unprecedented dangers of nuclear weapons. Since then, the conferences (attended by scientists from both West and East) have continued each year to discuss problems of disarmament, even during periods of acute international tension. Pugwash has always preferred behind-the-scenes diplomacy to overt pressure politics, and it is therefore difficult to assess its positive achievements. It certainly played a part in bringing about the Partial Test Ban Treaty and the Non-proliferation Treaty, and helped to initiate the Paris peace talks on Vietnam. However, it has failed conspicuously to attract strong support from the cadres of working scientists, particularly the younger age-groups, so that today its effectiveness is increasingly doubtful.

In many ways, Pugwash and SESPA represent extremes of the

spectrum of organisations set up to provoke scientists into awareness of their social responsibilities. Pugwash is run by elderly scientists, most of them distinguished members of 'the establishment'. SESPA consists largely of younger people and is anti-establishment. Pugwash works by quiet diplomacy in the corridors of power (so it believes), while SESPA groups throughout the United States hold protest meetings and marches, distribute leaflets and a gaudy magazine *Science for the People*, and use all the aggressive techniques of activist politics. (While Pugwash was being talked of as a potential recipient for a Nobel Peace Prize a few years back, SESPA was awarding the first of its annual M. F. Strangelove Awards 'for outstanding contributions to the modern theory of genocide and mass destruction' to Dr Michael May, director of the Lawrence Radiation Laboratory at Livermore, California, and Dr Edward Teller, a former Livermore director and 'father of the H-bomb'.) Though it has broadened its scope away from its original concern for nuclear weapons, Pugwash is devoted entirely to arms control and disarmament (with little more than lip service to the problems of the Third World and development). SESPA concerns itself with the whole of science and the whole of society – from community health care to automobile pollution.

In SESPA and similar organisations which have sprung up in recent years, scientific matters have become totally intertwined with those of social organisation, capitalist economics, the Vietnam war, and so on. Indeed, the 'New Critics' in American science – as the Stanford physicist Professor Martin Perl calls them[4] – see any attempt to separate such issues as entirely artificial. Whereas the normal response of the scientific community is to look for technical solutions to the evils of pollution, for example, the New Critics confront such wider questions as: What is wrong with the political environment which allows widespread pollution? What political means can be employed to improve matters? And what is the responsibility of the scientific community? 'Deeply embedded in their efforts is a new and critical examination of the political relationships between the scientific community, the universities, and the national government,' writes Perl. 'The New Critics have added the crucial ingredients of political understanding, political responsibility, and political action, to problems formerly considered as mainly technological concerns of the scientist.'

The campaign mounted by US scientists during 1969 to persuade the government not to deploy an anti-ballistic missile (ABM) system illustrates the new mood. It began with small groups of scientists and non-scientists in Boston, Detroit, Chicago, and Seattle, but grew quickly into a national crusade. ABM deployment was opposed on three principal grounds. At a cost of several billions of dollars, it would be a vast waste of human and physical resources sorely needed for constructive purposes. Secondly, it would decrease rather than increase national security and would impair international peace by initiating a new cycle of the nuclear arms race. Thirdly, it would strengthen and extend the power of the military industrial complex over society. (Many expert critics had strong doubts as to whether the system was even workable.) Although their advice is classified, it is thought that the majority of the President's own science advisers opposed ABM deployment. But more effective than any such behind-the-scenes opposition was the public campaign, which built up into an alliance between bodies such as SESPA, independent scientist critics, and concerned senators. Tactics included visits to sixty senators by SESPA activists during a meeting of the American Physical Society in Washington DC, a nationally televised anti-ABM press conference with a dozen congressmen, and an anti-ABM protest march of several hundred physicists to the White House. With the change from the Johnson to the Nixon administration, the opposition became even stronger and the President decided to replace the proposed Sentinel ABM system by the smaller and less provocative Safeguard system. Although the critics won only a token victory, therefore, what was remarkable was the extent to which the grass-roots critics were able to fight a battle on virtually equal terms with the President. At the final vote, the campaign to abandon the ABM system failed by just one vote to win senate approval.

This was the first time that a major military technological decision of an American administration had been strongly challenged. Hence to another target for critical attention – the closed channels through which governments receive technical information and advice from the scientific establishment. The critics see as unsatisfactory a system in which the establishment has triple roles of science adviser, lobbyist for government funds for science, and overseer of how those funds should be distributed and spent. 'To effectively obtain funds from the federal government,' Perl writes, 'the scientists must be able to convince the government of the value of the expenditure of these funds for science. To do this they must maintain reasonably friendly relations with both the legislative and executive branches of the government. If those same scientists are severely and publicly critical of the government's technological policies, it will be harder to obtain those funds.' Moreover, the nature of the issues referred to science advisers, the actual material presented, and the advice given by them, is usually private and privileged. In the USA, such material is released only rarely to Congress, the Press, or public, being kept secret not only for reasons of military security, but also on political grounds. An American president who goes ahead with a new technological project against the recommendations in an advisory report, for example, is likely to keep that report confidential. Despite scientific opposition, therefore, an unwise decision may gain support because of the apparent acquiescence of scientists in the decision-making.

Digging out information of this sort is extremely difficult, but some successes have been registered in recent years. One of the most interesting was in June 1971, when another critical organisation, the Scientists' Institute for Public Information (SIPI) took successful legal action to compel the US Atomic Energy Commission to disclose detailed information about the environmental effects of its fast breeder reactors before proceeding with their further development. Fast breeder reactors, while offering a much-needed source of energy in the years ahead as fossil fuels are depleted, will create considerable problems of disposing of high-level radioactive waste materials (see also p. 209). While some activist groups were so opposed to fast breeders on environmental grounds as to try to kill the entire development programme, SIPI was concerned simply to have the fullest possible information on which to make a balanced assessment. President Nixon had started the ball rolling in a special urgent message to Congress in which he requested some 100 million dollars to investigate new energy sources, with particular emphasis on the need for prompt financing of the development of fast-breeder reactors. SIPI feared that, in the headlong rush, environmental considerations were being played down, and resorted to the courts to force the AEC to release the relevant data.

Another American organisation, the Environmental Defense Fund, has specialised in using the law courts to sue polluters and other environmental destroyers.[5] It began in 1967, after a group of Long Island scientists and conservationists entered a suit aimed at preventing further use of DDT by the County Mosquito Control Commission. This followed observations of innumerable destructive effects – particularly the rapid decline of the osprey population on Gardiner's Island and Long Island, and the disappearance of crabs from local bays. The activists, who formed the nucleus of the new EDF, won their case, and further DDT use was banned. Since then, the EDF has entered suits involving such problems as atmospheric pollution from a pulp mill, the dumping of nerve gas by the US Army, and the degradation of rivers by damming, 'channelisation' and dredging. In work of this type, highly competent scientists and skilled attorneys are both vitally necessary.

One method widely used by the activists to open up the socio-scientific debate among scientists is to persuade scientific societies to mount social and political discussions as an integral part of their regular meetings. For several years now, the American Physical Society, which represents most of the professional physicists in the United States, has been polarised by such demands. In 1968, Professor Charles Schwartz from Berkeley introduced a motion calling for the inclusion of social issues on meeting agendas. The members rejected the motion but 30 per cent of those who voted were in favour. The opponents argued that science was neutral and that political motions should have no place in their affairs. More recently, in April 1972, Professor Robert March of the University of Wisconsin proposed a motion (backed by 275 other members of the APS) to graft crucial new aims on to the society's charter. To the words: 'the object of the society shall be the advancement and diffusion of the knowledge of physics', he proposed the following addition: '. . . in order to increase man's understanding of nature and to contribute to the enhancement of the quality of life for all people. The Society shall assist its members in the pursuit of these human goals, and it shall shun those activities which are judged to contribute harmfully to the welfare of mankind.'

March's motion did not succeed but the APS, prodded by earlier initiatives from Schwartz and others, has set up a continuing Forum on Physics and Society. Even this new section ran into trouble when it held its first symposium in April 1972, on 'Physicists and the Vietnam War'. Although the Forum was part of the formal proceedings of the APS, the executive secretary, Dr W. Havens of Cornell University, considered that the papers given dealt with physicists rather than physics, and that their abstracts therefore should not be printed in the *Bulletin of the American Physical Society* alongside those of scientific communications given at the meeting.[6] The battle goes on.

One successful attempt to raise an important political issue during what was nominally a purely scientific meeting took place at the International Congress for Microbiology, held in Mexico City in August 1970, when a resolution against work on biological warfare was debated and passed. The prime mover behind the resolution was Professor Carl-Göran Hedén, a microbiologist working at the Stockholm International Peace Research Institute (SIPRI). Apart from the need for biological disarmament *per se*, Hedén was passionately concerned about the redeployment of CBW resources for peaceful purposes. Speaking at the Third Conference on the Global Impact of Applied Microbiology, held in Bombay in December 1969, he listed the innumerable world problems requiring microbiological expertise – including the perfection of organisms to attack and control plant pests, and the improvement of methods for tackling epidemics – and then posed some pertinent questions about our capacity to solve them.

'Where do we find the best facilities for large-scale production of pathogens which destroy our biological resources?' he asked. 'Who are the experts in the aerosol delivery of such agents and the techniques available for stabilising them? Who have the facilities and the competence for producing large quantities of veterinary vaccines? Where can we learn about diagnostic procedures suitable for field work?' The answer was, of course, that by far the biggest concentrations of applied microbiologists at that time were to be found in government laboratories concerned with CBW. In the USA, some 3000 people, including 430 medical doctors, were working on CBW research at Fort Detrick. Another 1700 at Pine Bluff Arsenal, Arkansas (using equipment valued in 1966 at 138 million dollars) were producing the country's CBW munitions, which were field tested at the Dugway proving grounds in Utah, where another 1600 personnel worked. The Soviet Union, presumably, had comparable numbers involved in the same type of work. In contrast, the World Health Organisation at that time had a total staff of just 3500. Clearly, the redeployment of CBW expertise and resources could make an enormous impact on human welfare.

Against substantial opposition within the International Association of Microbiological Societies, which organised the Mexico City congress, Professor Hedén managed to place an anti-CBW resolution on the agenda for discussion. Once passed – after several hours of debate – the resolution became the expression of current opinion among the representatives of more than 50,000 microbiologists throughout the world. As such, it undoubtedly helped to swell political pressure for CBW disarmament which was then growing, and which culminated in the Convention on Biological Weapons signed by the great powers early in 1972. Since then, there has been some progress towards freeing for peaceful purposes the resources of manpower and equipment previously tied up in biological weapons development. In Britain, for example, the Microbiological Research Establishment at Porton Down, centre of much criticism from activist groups in the 1960s, has changed direction markedly towards greater involvement in public health and other civil work and in June, 1972, Fort Detrick was turned over to cancer research. (Although Porton never officially admitted to an offensive capability, it is scarcely possible to develop defensive methods against biological warfare without simultaneously gaining knowledge of use in weapons production.) Looking to the future, Professor Hedén now hopes that the United Nations will agree to create an International Microbiological Agency, charged with applying microbiology, for positive work in medicine, agriculture and allied fields.

Dissent can, of course, be costly. For Franklin Long, opposition to ABM deployment meant the withdrawal by Nixon in April 1969 of the post previously offered to Long as director of the National Science Foundation. For Charles Schwartz, it meant an end to research funds which he formerly received from the US Air Force for his esoteric research into atomic structure. In 1970, after the US Senate passed the Mansfield amendment, which ensured that all scientific research performed with support from the military must be related to the goals of the Department of Defense, Schwartz wrote to ask how the Air Force justified his research. 'Be assured that your work is vital to the aerospace mission of the Air Force,' said the reply from the D of D scientific officer. It went on to point out that security precautions prevented disclosure of why the work was vital. Schwartz felt that he could not comply with this state of affairs and terminated the contract.[7]

As well as the refusal of money from suspect sources on conscientious grounds, funds for dissenters and dissenting organisations often prove difficult to raise, and can dry up for political reasons. There was a classic example of this during 1971. Two US groups applied to the National Science Foundation for financial support. They were Science Service, highly respectable publishers of an uncritical and unexciting science magazine for laymen; and the Committee for Environmental Information (CEI), run by Professor Barry Commoner and other socially-concerned scientists. The CEI publishes the magazine *Environment*, which is devoted largely to thorough investigations of environmental pollution and allied matters. Science Service received 300,000 dollars; CEI, which had asked for 90,000 dollars, received nothing. The reason given was that it was not the NSF's role to support activities related towards applications of science in 'particular problem areas' – yet the Foundation was, at the same time, supporting a newsletter produced by the Southern Interstate Nuclear Board, founded to promote the development of nuclear power. Clearly, the NSF considered some specific applications of science more worthy than others.

British Society for Social Responsibility in Science

A group whose development illustrates many of the problems – financial difficulties included – which beset radical activist groups is the British Society for Social Responsibility in Science (BSSRS). Unlike its American counterpart founded in 1949, the Society for Social Responsibility in Science, which has members in 25 countries, the BSSRS is confined to one country. It was launched at a meeting in the Royal Society's rooms in London on 19 April,

1969, and was the brain child of Dr Steven Rose, a young biochemist then working at Imperial College, London, and his wife Hilary Rose, a sociologist at the London School of Economics. They were associated with a small band of young radical scientists, and supported by an impressive list of over forty Fellows of the Royal Society. In its earliest literature, the BSSRS stressed that 'the development of science is not predetermined but should depend upon the social choices of the community and the individual choices of the scientist'. In furtherance of this belief, the Society laid down four aims:

1. To stimulate amongst scientists an awareness of the social significance of science and of their corresponding social responsibilities both individually and collectively.
2. To draw the attention of all to the political, social and economic pressures affecting the development of science.
3. To draw public attention to the implications and consequences of scientific development and thus to create an informed public which could exercise choice in these matters.
4. To seek international exchange on these matters with similar groups in other countries.

By producing a regular news bulletin, holding meetings, and encouraging the formation of local groups throughout Britain, the Society aimed to take these objectives further. In particular, it would seek to identify critical developments likely to have a major effect on human life and the environment, conduct research into such effects, and inform the scientific community, the public and government. It would try to influence public policy on such matters when the actual or potential consequences of existing policy appeared 'undesirable', as well as urging the reform of science education at all levels, with the hope of encouraging social awareness and responsibility amongst future scientists.

The first problem for the fledgling BSSRS – common to all leftward organisations – was that the understanding of different supporters regarding these objectives differed markedly. For some – distinguished Fellows in particular – they were little more than glowing words. For others, they were a call to political activism. And even among the more politically active members of the Society, there were inbuilt contradictions and tensions, between middle-of-the-road liberals and committed Marxists and other sectarian socialists. As the months went by, these divergences became more obvious. While most supporters (even the distinguished Fellows) could applaud the BSSRS's star-studded international meeting on the Social Impact of Modern Biology,[8] held for three days in November 1970, there were some who felt that the Society should take a more aggressive stand over such issues as the use of CS gas to quell civil disturbances in Northern Ireland. When the Cumberland County Council approached the BSSRS for advice about the discharge of increased amounts of radioactive waste from the Atomic Energy Authority plant at Windscale, some members felt angry about the conclusions reached by other members who served on the working party examining the matter. The working party, consisting of Professor Eric Burhop and other senior scientists together with a number of younger physicists, concluded that the proposed increase was probably acceptably safe, but suggested further research which should be conducted if and when the UK AEA contemplated any further increase in the amount of discharge. Some members disputed this sensible, balanced advice, and felt instead that the Society should be committed firmly against *any* increase in environmental pollution.

Despite such internal troubles, the BSSRS grew in numerical strength. With a secretariat supported by the Rowntree Trust, it also grew in the impact it made, via the press, on public discussion and awareness of socio-scientific issues. Many of the original supporters melted away, and even the prime movers behind the formation of the Society, Steven and Hilary Rose, left in April 1971, pleading that the Society had 'vacillated interminably on many issues' and failed to provide a clear line and adequate leadership. Yet, with a membership of about 1200 by April 1972, and fifteen local groups, the range of the Society's work was impressive. It included major conferences on such topics as uranium enrichment and social factors in health and disease; submission of evidence to the Himsworth committee on the toxicity of CS gas; and campaigns on such subjects as the mental effects of the interrogation techniques used on detainees in Northern Ireland, and on race and intelligence.

This latter led to the suggestion by some radical scientists that, in view of the politically and socially evil ends to which the results might be put, there should be a moratorium on all research on intelligence in different racial groups (see p. 230). There are echoes here of the letter which Leo Szilard sent in February 1939 to Professor J. F. Joliot-Curie in France proposing that atomic physicists should agree voluntarily not to publish any new findings on the fission of uranium. Another idea thrown up by the BSSRS in its early days was that of the need to create an international network of 'spies for peace' – socially-responsible research scientists who would report on potentially dangerous work being pursued by their colleagues.

The event which received the widest publicity in the early months of the BSSRS was the Society's intervention at the annual meeting of the British Association for the Advancement of Science (BA), held in Durham in autumn, 1970. Despite the aims of its original founders, the BA became infamous, during the 1960s, for the scant attention paid, at its huge annual jamboree, to the social implications of science. Supported largely by schoolchildren, teachers, and elderly academics, the yearly meeting had become almost a caricature of itself, an annual binge in which charabanc tours and concerts always threatened to dwarf the scientific proceedings. That, at least, is how the BSSRS saw its benevolent and anachronistic target. 'Science is in crisis. Who would suspect it, from reading the BA programme?' asked a hand-out leaflet prepared by the BSSRS before the 1970 meeting. 'Where is the concern for pollution? Hidden in a Sunday afternoon tour of the filthy Durham coastline? Where is the concern about war? Will it be expressed by the Ministry of Defence representatives who address the meeting? . . . Where is the concern about priorities? British science has been warned by the State: serve industry or starve. Are our universities to educate scientists, or to produce units of scientific manpower? . . . Science cannot be transformed from within, only by a transformation of society. But the responsibility for this cannot be left to others. We are all responsible.'

The methods used to get the message across included a lengthy teach-in entitled 'Science is not neutral', attempts to raise social issues during scientific lectures, and the distribution of a series of daily leaflets commenting on the previous day's proceedings. At the various sectional meetings, BA members associated with the

BSSRS tried to raise questions about the social context of research papers related to guerrilla, nuclear and missile warfare. One speaker discussing infra-red devices, for example, was asked about his attitude to their use in night warfare in Vietnam. In most cases, chairmen stifled the questioning on the ground of irrelevance. The BSSRS responded by quoting the BA's own constitution to the effect that its meetings were 'designed to reflect, through discussion, the impact of science and technology on society', and argued that opportunities to discuss the social as well as the technical problems raised in the papers should be provided at all sessions. When this was refused, Society spokesmen called a hasty press conference announcing that they were pulling out, and departed in a petulant huff.

All this, of course, was strange indeed to the typical BA participant. Oddest of all were the events surrounding the opening session in Durham Cathedral at which the president, Lord Todd, gave his inaugural address entitled 'A time to think'. At the entrance to the cathedral, BSSRS members handed out copies of Todd's speech, annotated with extra comments highly critical of Todd's views of history, society, technology, and education. ('Advancing technology has brought with it great material benefits, which have been spread widely throughout the entire population,' Todd's speech read. 'Tell that to the Child Poverty Action Group' said the marginal note . . . and so on.) Afterwards, as members left the Cathedral, there was another surprise just outside on Palace Green, where a theatre group performed a 'nerve gas charade'. Actors writhed on the floor crying 'Let us breathe', while another spoke through a megaphone about the use of CS in Ulster. Emerging from the Cathedral in their academic robes, many BA regulars found the demonstration both unexpected and embarrassing. Not so the Bishop of Durham, Ian Ramsey, who, in a sermon preached the following Sunday, praised the BSSRS intervention as possibly more significant for both science and theology than the historic Huxley-Wilberforce confrontation over evolution at the BA meeting in 1860. Later in the week, participants at the BSSRS teach-in adopted the 'Durham resolution', which stated: 'As socially responsible scientists, we hereby undertake not to conceal from the public any information about the general nature of our research and about the dangerous uses to which it might be put.' Again, internal disagreements eventually won the day, as the resolution was defeated when put to the Society's AGM just a few weeks later.

Another stormy meeting held by the BSSRS took place at the Cambridge Union Society in July 1970. It was stimulated by Professor Arthur Jensen's publication, the previous winter, of a paper asking the question 'How much can we boost IQ and scholastic achievement?'[9] In his paper Jensen attributed the persistent IQ differences between US blacks and whites predominantly to differences in genetic constitution, and questioned the value of educational compensatory programmes. The meeting itself was organised as a forum at which Jensen confronted several passionate critics, who attacked both the reliability and significance of IQ tests and stressed the wider implications of such work.[10] Dr Martin Richards, for example, argued that, quite apart from what it aimed to measure, the IQ test was a political instrument. Jensen stood up well to the sustained criticism, but given the style and format of the meeting, its outcome – massive rejection of the Jensen position – was a foregone conclusion.

The BSSRS has still not solved its inner conflict of objectives and tactics. It may never do so – and may, indeed, never wish to. Typical of the internal tensions were two conflicting proposals about future policies put to the annual general meeting in 1971. One idea was that the Society should become little more than an agency for distributing independent, authoritative information about scientific issues – a worthy and necessary role, though a more passive one than the organisers had in mind in the early days. The other was a move towards explicit alignment with revolutionary and working class movements. The Society found itself split down the middle, but averted schism by assembling a compromise solution consisting of the following four propositions:

1. The Society should take as a priority to provide scientific expertise and advice to those groups in society which do not normally have access to it.

2. The Society should continue to comment on current political issues, though avoiding any party political alignment.

3. The current ideological debate should continue.

4. Particular groups within the Society should not be restricted by this consensus, but only by the BSSRS constitution.

Discussing these proposals early in 1972, BSSRS chairman Jonathan Rosenhead pointed out that they were simply ground rules for operations until such time as the debate moved on far enough for a new area of agreement to emerge.[11] 'But the most important task for all of us is to continue arguing and acting the issues out – where we work, where we live, in the press, inside the BSSRS, and inside our professional organisations,' he wrote. 'Only in this way can we hope to educate ourselves, and in the process help to create a public which can understand the issues involved in trying to control its science.'

One-man campaigns

One certain way to avoid internal organisational conflicts is to remain a one-man campaigner. Björn Gillberg, a geneticist at the State Agricultural College in Uppsala, Sweden, is one such. After beginning as a student of educational psychology, Gillberg turned to genetics and molecular biology, at which point he became interested in possibly unrecognised environmental hazards of chemicals which are capable of causing genetic mutations in man. He wrote a book about potential mutagens used as pesticides, food additives, and drugs, which was published in 1969, and today he is a nationally known figure in Sweden. Housewives, for example, avoid supermarkets which sell the products Gillberg has condemned.[12] He sees one of his major roles as providing scientific information to the public and particular groups lacking access to specialist data, and works in collaboration with SIPI in the United States.

Gillberg too has had to face financial problems as a direct result of his activities. In 1971, he wrote a second book in which he criticised Swedish policy and regulations regarding detergents – particularly the use of optical brighteners, which can cause cancer in laboratory animals and which make clothes look whiter but have no cleansing action. In many respects, co-operation between Swedish industry and the government (see p. 216) has worked well in containing potential environmental problems. But, Gillberg argues, there has been a price to pay in surrendering some independence in regulatory authorities. The state Environment Protection Board, for example, set up a 'detergent committee', on which detergent manufacturers' representatives held the majority of positions. Gillberg attacked this set-up and various aspects of current legislation in his book, and shortly afterwards the state Board for Technical Development announced that it could no longer support his research work on nitrogen-fixing bacteria. This could have been coincidental, but Gillberg believed otherwise. More likely, it was an attempt to silence him. Months later, after extensive public debate over the incident, many thousands of Swedish crowns had come in from the public to support Gillberg's work. Almost as a bonus, the Swedish International Development Agency announced that it would finance his research in future. By that time, Gillberg had become probably the most famous scientist in Sweden.

Several other scientists who have been prominent in battering their fellow scientists and the general public into responsible concern and action over socio-scientific issues have been environmentalists. Dr Paul Ehrlich,* professor of biology at Stanford University, and founder of Zero Population Growth (a grass-roots birth and population control organisation) is probably the best known. Ehrlich is chiefly concerned to warn about the dangers of population growth – which he believes is outstripping food production, the Green Revolution notwithstanding – and relates all other dimensions of the environmental crisis to population pressure. 'While you are reading these words, five people, mostly children, have died of starvation – and forty more babies have been born,' he warns on the cover of his best-selling book *The Population Bomb*,[13] first published in 1968. In it, he lists man's fifteen 'inalienable rights', which can be bought at the cost of giving up the right to irresponsible reproduction. They are the right to limit our families, to eat (and to eat meat), to drink pure water, to live uncrowded in decent homes, to avoid regimentation, to hunt and fish and view natural beauty, to breathe clean air, to avoid pesticide poisoning and thermonuclear war, to educate our children, to have grandchildren and great-grandchildren – and to have silence if we want it.

* Not related in any way to the Paul Ehrlich mentioned on p. 111.

In his quest, Ehrlich shows prodigious energy, giving some 100 public lectures and 200 radio and television broadcasts each year. He also exaggerates his case, saying in effect that over-statement is safer than under-statement when faced with such mammoth problems. Typical of his technique was his speech to Britain's Institute of Biology, held in London in September 1969, at which he forecast worldwide plagues, thermonuclear war, overwhelming pollution, ecological catastrophe, and the virtual collapse of Britain by the end of the century. 'If current trends continue,' he said, 'by the year 2000 the United Kingdom will simply be a small group of impoverished islands, inhabited by some 70 million hungry people, of little or no concern to the other 5–7 billion inhabitants of a sick world.' After articulating a clutch of memorable phrases ('siren song of the myopic optimists', 'tidal waves sweeping the country', 'political turmoil will rule'), he ended by warning: 'If I were a gambler, I would take even money that England will not exist in the year 2000.'[14]

The crusades of Barry Commoner, director of the Center for the Biology of Natural Systems at Washington University, St Louis, and chairman of the Board of Directors of SIPI, are almost as widely known as those of Paul Ehrlich. But Commoner differs in his basic diagnosis of man's predicament. He believes that both pollution in rich countries and overpopulation in the Third World are results of the exploitation of technological, economic, and political power by a private enterprise system that is greedy for profit. In his book *The Closing Circle*,[15] he argues that new technologies, driven by the profit motive, have brought man increasingly outside the balance of nature. For example, as synthetic fibres have replaced cotton and wool, aluminium and plastic replaced wood, and artificial fertilisers replaced manure, production of the new materials has led to greatly increased pollution. Under neither capitalism nor communism does the economic system take full account of the social costs of the goods it produces. The environmental crisis is not, therefore, a mere biological or technological problem; it must be seen in its social context. Man must mend his ways. Commoner was also the founder (in 1958) of the St Louis Committee for Atomic Information, which set out to ensure that the US public played an active part in inquiries into the siting of nuclear power stations.

Later, the organisation became the Committee for Environmental Information, which today publishes *Environment* magazine containing carefully-researched inquiries into environmental problems and attacks on government departments and private companies. In 1971, for example, the CEI helped to show that the large quantities of nerve gas which the US Army wished to dump in the Atlantic ocean could instead be safely detoxified on the spot.

Barry Commoner was also responsible for introducing the concept of 'critical science',[16] which entails the identification and analysis of damage caused by rampant technology and *laissez-faire* science, together with public criticism and campaigning for corrective action. Rather than scientists leaving their laboratories once in a while to ponder the social and environmental problems they have created, critical science involves a total approach in which science, and political and social action, possibly inconvenient for the interests involved, are virtually inseparable. 'The work of inquiry is largely futile unless it is followed up by exposure and campaigning; and hence critical science is inevitably and essentially political,' writes Dr Jerry Ravetz.[17] 'Its style of politics is not that of the modern mass movements or even that of "pressure groups" representing a particular constituency with a distinct set of interests; it is more like the politics of the Enlightenment, where a small minority uses reason, argument, and a mixture of political tactics to arouse public concern on matters of human welfare.' Critical science is as yet at an early stage of development, but two recent campaigns conducted in that spirit were Professor Ernest Sternglass's attempt to convince the scientific community and the public of the harmful effects of low level radiation from nuclear bomb tests[18] and the warnings of Drs John Gofman and Arthur Tamplin about alleged radiation dangers from nuclear reactors.[19] Former employees of the Atomic Energy Commission, Gofman and Tamplin believe that reactor safety standards are too lax, and that the present closed decision-making system should be replaced by more open involvement of the public and the scientific community.

The scientist who comes closest to the total distrust of science and the world it has created voiced by outside critics such as Theodore Roszak is the distinguished bacteriologist René Dubos. In his book, *Reason Awake*,[20] he discusses the contrast between the past, when man was threatened by natural forces he could not control, and the present day, when the most potent fears derive directly from science and technology. He emphasises the many ways in which scientific progress has created new dangers and threats to human welfare and argues for the total integration of all fields of science in the service of humanity. For Dubos, this *does not* mean rejuvenating the utopian optimism felt towards science in the pre-war years; it demands a much more critical temper, a change in society's attitude towards science and the scientist's own conception of his role and responsibility.

What should be the research worker's attitude when he discovers something which can be put to evil ends? One answer comes from the Yale University biologist, Dr Arthur Galston. In the early 1940s, he discovered, and published reports about, a chemical that could increase the yield of soyabean by 30 per cent per acre. Later, scientists at Fort Detrick found that the same and related chemicals could also be used for defoliation. The American forces tried them out experimentally in the Korean war, and then used them massively in Vietnam (see p. 84). 'My findings helped in the development of this technology,' Galston said some years later at a meeting on the social responsibility of the scientist. 'Should I feel guilty about it? Should I get involved?'[21] Galston's own answer has been to prod his colleagues into studying the ecological and public health effects of herbicides, and to compel the US government to reveal the truth about their use – culminating in Nixon's phasing out of defoliation in 1970. 'Out of my social background, out of what ethics had been built into me personally. . . . I felt moved to pursue further the implications of what I had done,' Galston said. 'Each individual must find in his own background, within the relevance of the society in which he is placed, the rules that tell him about social involvement.'

An altogether different response to his own work was that of Jim Shapiro, a member of the team at Harvard Medical School which, in November 1969, announced the isolation of a gene, part of the hereditary material of a bacterium.[22] At a press conference called to publicise the discovery, Shapiro and his colleagues spoke about the potential significance of their work in making feasible genetic engineering. As a means of tinkering with human heredity, this as yet theoretical technique could be used to treat disease, or it could be turned to more sinister ends in modifying human behaviour, and the Harvard group felt obliged to warn the public of the dangerous potential of what they had done. In a letter to *Nature*[23] shortly afterwards, three of the team answered critics who had criticised their attitude as alarmist. They quoted Los Angeles smog, DDT in milk, Vietnam, Hiroshima, and other examples of 'the uncontrolled use of science by governments and private corporations' to justify their pessimism, and called for scientists to work towards radical political change. For Jim Shapiro, the crunch came with a telephone call from a large private medical foundation asking him to join a secret and well financed crash programme to make genetic engineering a reality within a few years. 'That telephone call shocked me,' Shapiro said.[24] 'It shows that an élite group of rich men and complacent scientists are ready to rush ahead with a branch of biology which presents society with the gravest moral and political problems, and to do it in secret, concealing from the public the very facts it is essential that they know.' Shortly afterwards, Shapiro abandoned his highly promising career in science and devoted himself to full-time activist politics. More recently, another member of the team, Jon Beckwith, accepting the Eli Lilly Award at the AGM of the American Society for Microbiology, announced that he was so repelled by the self-interested policies of money-hungry drug companies that he was donating the money to the Black Panther Free Health Movement.

Another motive for 'dropping out' from the scientific world is typified by the case of Robin Clarke. A qualified scientist, he became editor of *Science Journal* in Britain and later went to work for UNESCO in Paris. At the end of 1972 he left to set up a 'soft technology research community' in Wales. 'Technology should be valid for all men at all time,' Clarke writes.[25] Specifically, this excludes 'all technologies which depend on the manipulation of one group of men by another group; all the technologies which if practised globally would have undesirable cumulative effects; and all the technologies which depend on using resources in such a way that they will not be available for future generations. It is, in other words, a formulation which recognises . . . that we now need new technologies which neither pollute the planet nor mortgage the future for unborn generations by using and exhausting irreplaceable resources.' The accompanying Table contrasts 36 'utopian

Utopian characteristics of soft (alternative) technology

	Hard Technology Society	*Soft Technology Society*
1	Ecologically unsound	Ecologically sound
2	Large energy input	Small energy input
3	High pollution rate	Low pollution rate
4	'one way' use of materials and energy	reversible materials and renewable energy sources only
5	functional for limited time only	functional for all time
6	mass production	craft industry
7	high specialisation	low specialisation
8	nuclear family	communal units
9	city emphasis	village emphasis
10	consensus politics	democratic politics
11	technical boundaries set by wealth	technical boundaries set by nature
12	alienation from nature	integration with nature
13	world-wide trade	local bartering
14	destructive of local culture	compatible with local culture
15	technology liable to misuse	safeguards against misuse
16	highly destructive of other species	dependent on well-being of other species
17	innovation regulated by profit and war	innovation regulated by need
18	growth orientated economy	steady state economy
19	capital intensive	labour intensive
20	centralist	decentralist
21	alienates young and old	integrates young and old
22	general efficiency increases with size	general efficiency increases with smallness
23	operating modes too complicated for general comprehension	operating modes understandable by all
24	technological accidents frequent and serious	technological accidents few and unimportant
25	singular solutions to technical and social problems	diverse solutions to technical and social problems
26	agricultural emphasis on monoculture	agricultural emphasis on diversity
27	quantity criteria highly valued	quality criteria highly valued
28	work undertaken primarily for income	work undertaken primarily for satisfaction

	Hard Technology Society	*Soft Technology Society*
29	food production by specialised industry	food production shared by all
30	science and technology alienated from culture	science and technology integrated with culture
31	small units totally dependent on others	small units self-sufficient
32	science and technology performed by specialist élites	science and technology performed by all
33	science and technology divorced from other forms of knowledge	science and technology integrated with other forms of knowledge
34	strong work/leisure distinction	weak or non-existent work/leisure distinction
35	high unemployment	(concept not valid)
36	technical goals valid for only a small proportion of globe for a finite time	technical goals valid 'for all men for all time'

characteristics' of a society based on soft technology with the corresponding characteristics of a hard technology society. Clarke's aim is to show, by practical example, that soft technology – an attempt to reverse the spiral of profit and pollution discerned by Commoner – is actually feasible. Moreover, in place of conventional science, which is 'performed by a specialist élite who unconsciously determine the kind of life people must lead', soft science and technology 'would be carried out by those who need it for purposes defined by themselves and would never advance beyond the comprehension of those operating it.'

Finally, a proposal from a longstanding writer on science and society which the protagonists of critical science have both welcomed as a radical advance and condemned as exclusive and elitist. In July 1971, Dr Jacob Bronowski, the Director of the Council for Biology in Human Affairs at the Salk Institute, San Diego, suggested that science should become disestablished. It should dissociate itself urgently from the apparatus of government in general, and government grants and contracts in particular.[26] Bronowski quoted numerous factors to support his case – including the obligations on scientists and institutions dependent on military and quasi-military financial support (there must, for example, be no ban on armed services recruiting in universities receiving US government grants) and growing political pressures on scientists to conform (in October 1969 it was revealed that the National Institutes of Health, Bethesda, maintained secret lists of scientists considered suspect for political reasons and who must not therefore be appointed to committees). 'There is now a duty laid on scientists to set an incorruptible standard for public morality,' wrote Bronowski, pointing out that it was no longer tenable for scientists to adopt the stance which Robert Oppenheimer tried to maintain – 'to be a technical adviser on weapons on some days and an international conscience on others'. Hence the need for a complete separation between science and government in all countries. This would mean hardship for a time, but the eventual aim should be 'to put science in the hands of the scientists'. Having won financial independence, in the form of a single overall grant for research, the body of scientists should 'assign its own priorities on behalf of the community and with it, and divide the overall grant at its disposal accordingly'.

That, of course, raises the greatest question of all. It is the one we shall turn to in the final chapter.

Chapter 11
What is science for?

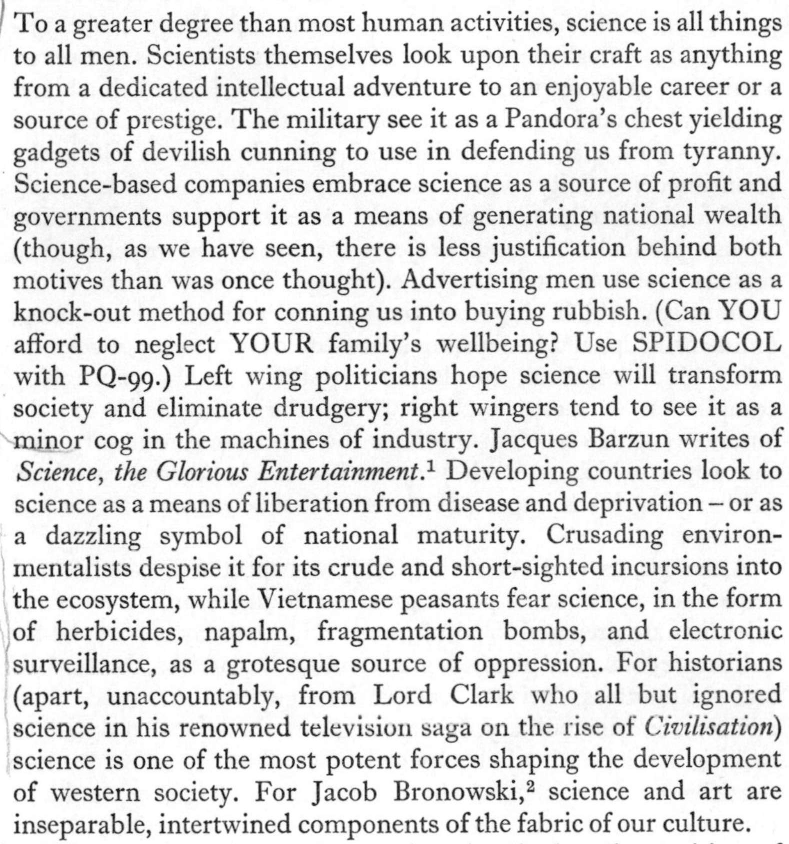

To a greater degree than most human activities, science is all things to all men. Scientists themselves look upon their craft as anything from a dedicated intellectual adventure to an enjoyable career or a source of prestige. The military see it as a Pandora's chest yielding gadgets of devilish cunning to use in defending us from tyranny. Science-based companies embrace science as a source of profit and governments support it as a means of generating national wealth (though, as we have seen, there is less justification behind both motives than was once thought). Advertising men use science as a knock-out method for conning us into buying rubbish. (Can YOU afford to neglect YOUR family's wellbeing? Use SPIDOCOL with PQ-99.) Left wing politicians hope science will transform society and eliminate drudgery; right wingers tend to see it as a minor cog in the machines of industry. Jacques Barzun writes of *Science, the Glorious Entertainment*.[1] Developing countries look to science as a means of liberation from disease and deprivation – or as a dazzling symbol of national maturity. Crusading environmentalists despise it for its crude and short-sighted incursions into the ecosystem, while Vietnamese peasants fear science, in the form of herbicides, napalm, fragmentation bombs, and electronic surveillance, as a grotesque source of oppression. For historians (apart, unaccountably, from Lord Clark who all but ignored science in his renowned television saga on the rise of *Civilisation*) science is one of the most potent forces shaping the development of western society. For Jacob Bronowski,[2] science and art are inseparable, intertwined components of the fabric of our culture.

What we must now ask ourselves is whether the position of science in society – the importance we attach to it, and the ways in which we decide how much and what sort of science to support – is the right one. Sir Brian Flowers[3] has written about two sets of forces acting in the affairs of science: external forces representing the aims of society, and internal forces representing 'the natural development of science'. Problems arise whenever this balance of forces breaks down. We should, for example, guard against a situation where the inner forces – by which Flowers means the internal logic which propels scientific experimentation and speculation forward – are so compelling that they push science beyond our immediate capacity to deal with either the social or practical problems it creates. Similarly, it is possible for an ill-informed public (or enthusiastic politicians on its behalf) to make unreasonable, unrealistic demands of scientists. There is, however, a third set of forces which we have encountered in this book and which we should consider in assessing the place of science in society – the social, commercial, and political pressures within the scientific community itself. Science is not driven onwards simply by objective intellectual analysis and experimental skill. Scientists choose particular research projects for social and psychological reasons as well as intellectual ones (p. 39). The same is true of the committees which dispense money for research. Politicking, grantsmanship, and in-fighting shape the development of science, every bit as much as the daring hypothesis or the painstaking compilation of meter readings. Particularly disquieting is the autonomy which so often insulates the scientific community, and the way in which research projects become bandwaggons whose further progress is important to individual scientists' careers and prestige.

Although, as we have seen (p. 29), unexpected discoveries and unplanned experiments play a greater role in research than many scientists are prepared to admit, most science today is pursued for deliberate ends. The planners are scientists themselves, their cabals and committees, private companies, governments and their advisers, and the military. In arguing, as I shall do, that decision-making in science should be opened-up at all levels and made more sensitive to the wishes of the community at large, it would be convenient to dissolve the entire problem in some exhaustive, overall philosophy. Discussing environmental crisis, for example, Hugh Montefiore, the Bishop of Kingston, argues that only when men come to believe that they 'hold dominion over nature as trustees and stewards of God', will they feel an inalienable duty towards the environment. Though his book *The Question Mark*:

The End of Homo Sapiens[4] is packed with technical information, Montefiore claims that religious belief offers 'the only hope for posterity', and he advocates a 'remythologised and revitalised Christianity' as the answer. The Marxist zoologist Harry Rothman, on the other hand, sees industrial pollution as an inevitable accompaniment of capitalism and prescribes revolution as his 'only hope'.[5] It is, of course, true that dilemmas such as those of environment and scientific priorities come back eventually to basic political and philosophical convictions, yet it is difficult to accept any such all-embracing analysis as 'the only' hope. As Dr Joseph Needham points out in a foreword to Montefiore's book, some of the most interesting early examples of conscious nature-protection can be traced to ancient China, where neither of the two religions, Confucianism nor Taoism, was theistic. Christianity does not have a particularly good record in ecology. On the other hand, the socialism of the Soviet Union has not been conspicuously successful in preventing or containing environmental pollution. For Rothman, such disastrous failures as the pollution of Russia's Lake Baikal (rivalling that of Lake Erie in the United States) have to be explained as results of the distortion of socialism by Soviet-style bureaucracy. There are many other untidy discrepancies on both sides. So it is with any single-minded panacea for the world's ills: death by a thousand qualifications.

In any case, whatever we believe as individuals, there is little point in advocating policies which, in a diverse and untidy world, have no chance of being widely accepted in the foreseeable future. While research is the art of the soluble, politics is the art of the possible.[6] Far from being a homogeneous juggernaut, as it is sometimes portrayed, modern society is a complex inferno of conflicting interests, and it is against this background that we must consider the place of science in society and the means of making changes in the way we support science and exploit its insights and end-products.

Another consideration should inform our thinking – the enormity of some of the problems created or about to be created by science and the extent to which 'the experts' disagree about them. The larger the problems, it seems, the more extreme the positions taken up by protagonists in opposing corners. The great doom debate, raging in recent years over the proposition that we face catastrophe as a result of over-population, depletion of resources, and global pollution, is the quintessential example. On one side are such assorted bedfellows as Sir Peter Medawar (who believes that our greatest danger is a failure of nerve – a lack of confidence that, unlike our predecessors in earlier times, we now have *time* to solve our problems);[7] John Maddox, former editor of *Nature* (who argues that the price mechanism and human ingenuity will take care of resource depletion; that intelligent, adaptable, *Homo sapiens* will conquer formidable obstacles in the future as in the past; and that there is no reason to fear science being turned to malevolent ends);[8] and politicians like Jeremy Bray (who places his faith in our capacity to cope by political action).[9] On the other side the doomsters include Ehrlich, Commoner, and Dr Dennis Meadows and his colleagues at the Massachusetts Institute of Technology. In 1972 Meadows's group published its computer study (part of the 'Project on the Predicament of Mankind', organised by the Club of Rome, an informal international association of scientists, industrialists and civil servants) which suggested that the world cannot support present rates of economic and population growth for more than a few decades.[10] The report has proved to be highly controversial. As Ehrlich and Commoner exemplify, even the diagnosis of doom, let alone prognosis, varies widely. On such major world problems, it often seems that optimism or pessimism is based largely on temperament, rather than on dependable fact.

There is similar disquieting uncertainty over many of the individual, defined problems stemming from our uses of science. We do not, for example, know for sure whether there is a 'safe limit' for exposure to nuclear radiation. Although studies of the survivors of Hiroshima and Nagasaki have yielded information about the effects of radiation, and though experiments on animals also provide relevant data, there is still controversy about what should be the maximum permissible radiation dose for human beings. The International Committee on Radiological Protection, which sets international standards, based its initial recommendations on the idea of a threshold – a level below which radiation doses were assumed to be entirely safe. Some scientists still accept this idea, but others – notably John Gofman and Arthur Tamplin – believe that the risk of damage is proportional to dose even at the lowest levels.[11] The dispute is of obvious public concern in the siting of nuclear power stations, the setting of safety standards against the possibility of accidental leakage of radioactive materials, and the continuing disposal of radioactive waste materials into the atmosphere and sea from nuclear power stations and fuel processing plants. Yet we simply do not have entirely reliable information on the central point of safety levels.

Despite this lacuna, nuclear power development goes ahead. And all is not well. In the early 1970s the US Atomic Energy Commission discovered that a crucially important safety device used in nuclear power reactors – the emergency core cooling system – was less reliable than had been thought previously,[12] justifying criticisms from environmentalists about safety standards. Moreover, the new breeder reactors now being developed throughout the world will yield large quantities of highly radioactive waste which will have to be stored away from the environment for thousands of years. One isotope produced will be dangerous for some 200,000 years. Yet the necessary storage methods have not yet been devised. As recently as May 1972, it was reported that America's AEC faced the possibility of receiving its first batches of this nuclear waste – one of the most poisonous materials the world has ever known – at the end of the 1970s and having nowhere to put them.

Another potential problem area which is the subject of wild disagreement among specialists is the feasibility of genetic engineering. Some scientists[13] believe that it will prove possible to ameliorate certain hereditary diseases by altering the genes inside human cells. Others fear that such techniques could be turned to evil ends. They could, for example, be used to produce several copies of the same person by 'cloning'. Identical Einsteins or identical Hitlers could be reared, starting with an ordinary body (not necessarily reproductive) cell of the person in question. Other medical scientists are much more sceptical of such possibilities. One of them, the distinguished immunologist Sir Macfarlane Burnet,[14] is dubious about *any* possible contributions of the science of molecular biology to medicine, even to such problems as cancer which others believe are most likely to succumb to this approach.

There is a long list of other possible scientific developments, all of them based on research now under way, all of them with enormous potential repercussions in society, yet all subject to conflicting expert opinion as to their technical feasibility and man's capacity to cope with any social problems they create. These include the possibility of mind control,[15] abortion pills, new techniques of communication, the conquest of ageing, the fertilisation of the human ovum outside the body and its reimplantation in the womb, and many more. Moreover, some of the sectors of science which will pose the greatest social dilemmas in five years time may well be those which today look the most innocent. They are also those which, as pointed out in Chapter 4, are largely in the hands of scientists rather than technologists. This is doubly disturbing because the science end of the science-technology spectrum is less sensitive to social pressures than is technology. The training of the research scientist is much narrower, he is more insulated from social pressures, yet his work is liable to have the most far-reaching consequences.

All of this suggests that vigilance and open appraisal should be the touchstones of our social approach to the control of science. And, if one must choose between the contemporary camps of optimism and pessimism towards the use and abuse of science, I fear one must opt for pessimism. We have learned not to accept the reassurances of experts about future developments, dangers, (or promises) of science. Lord Rutherford believed that splitting the atom would have no practical consequences. More recently, experts dismissed the early warnings of Rachel Carson (many of them later vindicated), and not so many years ago a British Astronomer-Royal ridiculed the idea of space travel as 'bilge'. Secondly, despite the gargantuan positive achievements of science, we have made ghastly errors in the past and risk worse mistakes in the future because of our greater dependence on science. Above all, we have to reckon with the increasing rate of technological change. As we have seen (p. 92), the accelerating pace at which laboratory discoveries reach practical fruition can be greatly exaggerated. What is true is that, at the point of application, the products and processes of science are deployed ever more quickly and extensively. A new synthetic chemical, for example, introduced as a pesticide, food additive, or detergent can be used by millions of people within a few weeks. Whatever tests have been conducted beforehand, every occasion on which this happens constitutes an experiment – an experiment on a massive scale. It is only a matter of time before a chemical deployed in this way is found to cause mutation or cancer.

Some scientists, many politicians, and most administrators believe that the mammoth problems of regulating modern science require the magisterial application of unified 'science policy', in which science is treated as something esoteric and isolated from the rest of society. Reports are issued and weighty meetings held at which we hear calls for a national or international science policy in which every crumb of scientific expenditure for the next five or ten years is neatly tabulated. Such efforts always fail, and we should be thankful. It is tempting to dream of a highly efficient overall plan for science, in which priorities are rationally ordered, but the diversity of modern society means that there is insufficient common ground to act as a basis for such policymaking. What we need instead is to see science as an integral part of social development, and to democratise science, both internally and externally, to render it more sensitive to the various strands and needs of society. We do not want to be dominated by 'a national science policy', any more than it is sensible to defuse creativity by establishing a state-owned pharmaceutical industry. International bodies and agencies are particularly adept at holding boring conferences on science policy, and it is therefore all the more welcome that in mid-1971 the Organisation for Economic Co-operation and Development published a report, since known as the Harvey Brooks report[16] after its chairman, which was not only healthily sceptical of the need for everlasting economic growth but which also discussed science as an integral part of human affairs rather than as a separate and special activity.

One of the most serious defects surrounding science throughout the world is the neurotic degree of secrecy which envelops so much decision-making – even when no considerations of military or national security are involved. President Kennedy's science advisers did not hear about the decision to begin the *Apollo* programme – ostensibly a scientific mission – until afterwards. The British people still do not know on whose advice their government decided not to join the CERN accelerator project in 1968. Countless other examples could be given. When, in 1969–71, independent, reputable scientific evidence cast serious doubts on the safety of a substance used in a fly-paper marketed in Britain, which acts in a similar way to nerve gases, the company concerned could not pass on its own toxicity results for independent appraisal in a scientific publication because it had given them in confidence to a government committee. The government department concerned could not divulge information presented to it confidentially.[17] Even when information *is* supposedly available, the procedures for gaining access to it can be so tedious and vulnerable to blocking action as to be almost opaque to inquiries. In the United States, for example, the Freedom of Information Act stipulates that every government agency shall, with certain specified exceptions, make its records promptly available to any person on request. In practice, numerous tactics of delay and evasion are employed to thwart the purpose of the Act, and have been used to block the release of information on environmental and other important para-scientific issues.[18] In Britain, although data about waste discharged into rivers, for example, is supposedly open to inquiry from people with a legitimate interest, the information is virtually impossible to obtain in many cases.[19]

Secrecy is one of the reasons why we, as citizens, have less formal influence over the course of scientific research and its practical deployment than over virtually any other aspect of modern society. The contrast can be seen most clearly if we consider ourselves as consumers. The individual's influence as a consumer – whether of food, clothes, magazines, or music – is considerable, assuming that he understands his power and learns how to use it. Even the most insidious and costly adventures of the ad-men and PR-operators come to naught when the citizen says no. And just as governments can mould events by using their buying policies, so the growth of consumer organisations in recent years has affected the quality and range of goods available to a degree that would have seemed impossible only ten years ago. Science, on the other hand, while it continuously generates a staggering range of *possible* future developments, is largely insensitive to the views of those outside the scientific community. The citizen *qua* citizen usually has no means of knowing about such choices at the right time or of influencing decisions on them.

Consequences of greater democratisation

It can, of course, be argued that the suggestion of greater public participation is naive, impractical, or idealistic. For the reasons we considered in Chapter 4, scientists tend to resent the idea. Why should bankers and bus conductors decide what biophysicists do with their time? How could they even begin to understand the complexities involved? Do they care anyway? The answer is threefold. First, it is now abundantly clear that the experts *do not* always know best; scientists have fathered so many obscenities and failed to foresee so many ill effects of their work that they have forfeited the automatic respect they once had as wise men in the community. Secondly, extrapolating the type of evidence collected in Chapter 8, it is likely that whatever the scientific community itself does or does not do to democratise decision-making, people are going to *demand* to know much more about what scientists are up to and why. 'In a democratic society,' says Alvin Weinberg, 'the public's right of access to the debate is as great as the public demands it to be.'[20] And people will be increasingly interested not only in the more obviously applied aspects of science, but also in the reasons why money is spent on new atom smashers or esoteric cell biology. Though an excellent development in itself, in line with growing demands for participation in other walks of life, a movement of this sort could pose a dangerous nuisance. It could be ill-informed and disruptive, gaining strength from those silly anarchistic movements in which expertise (particularly scientific expertise) is totally despised. Essentially anti-science ludditism of that complexion could be forestalled if we begin *now* to stimulate public awareness of the dilemmas and choices in science, and to fashion new instruments of democratic influence.

The most compelling reason in favour of greater involvement of the citizen in scientific policy-making is one of basic democratic right. Who but you and I and the other members of the community should decide between the many conflicts of financial and social cost and benefits arising from science? Our world and our well-being are at stake. Take the perplexing problem of cancer research. How much of the £5 million spent on cancer research every year in Britain, for example, supports outlandish biology which might – just might – yield clues to the riddle of cancer, and how much goes to improve the prospects for those now suffering from the disease? Dependable figures are not available, and the community at large has never been given an opportunity to make its views felt on such questions. Yet more than half of all cancer patients could be cured if the disease were diagnosed in its earliest stages and treated promptly.[21] Many tumours are *not* detected sufficiently early. And some dubious science is supported in the name of cancer research. If such facts were more widely known among the people who contribute to charitable research funds, and those who pay taxes that help to support research, there might well be a popular consensus that a greater proportion of resources be ploughed into diagnosis and treatment – and indeed into research specifically designed to improve methods of diagnosis and treatment. This is one example of an area where the public clearly has a real and legitimate concern not only with the *deployment* of science, in the form of medical care, but also with money spent on abstruse laboratory research.

There are enormous risks in arguing that decisions on such matters should be influenced by popular vote. Society rarely progresses on such a basis; in those countries which have abandoned capital punishment, for example, a referendum at the time would probably have found in favour of retaining it. With science, there is a particular danger of short-sightedness and a failure to understand the importance of speculative research as against the application of immediately useful techniques. For that reason, it is arguable whether the community at large should have a veto or even a casting vote in deciding research priorities. What we must try to do is create channels through which the pressure of the citizen can influence such decisions more directly. This requires that scientists explain their work and its consequences publicly and expose their decisions to democratic appraisal. It also places considerable responsibility on the press and other media in interpreting science and discussing critically its social implications. We do not want communal *control* over science, but a good deal more public *influence* – if only as a healthy corrective to the present autonomy and internal politicking of the scientific community, and the massive social and political influence of science-based corporations such as IBM. And if it be argued that fashion plays a capricious part in determining what people feel to be important (in 1965, 41 per cent of people questioned in Britain did not know whether the UK had an independent nuclear deterrent and 29 per cent thought it did not – just a few years after the deterrent was a subject of intense and passionate public debate),[22] then we must remember that fashion plays its capricious part within the scientific community too.

In his Trueman Wood lecture in December 1971, Lord Rothschild gave an excellent example of a conflict of priorities on the grand scale. The scientific community in Britain had been discussing the desirability of building a 'high flux beam reactor' (one of Alvin Weinberg's symbols of our age!), without which British science might begin to lag behind the rest of the scientific world. Yet the same money could buy: twenty miles of motorway, four hospitals, 300 adventure playgrounds for under-privileged children, four 1000-inmate prisons, 200,000 television sets for the old and housebound, two Jumbo jets, 4500 houses, or Titian's *Death of Actaeon* (at a tenth the cost). The dilemma of adjudicating between such priorities arises continuously in the competition for money within science and between the demands of science and other needs of society. The citizen could and should be made aware of such perplexities and be given more weight in resolving them.

As Steven and Hilary Rose argue in their book *Science and Society*,[23] to render decision-making about science more open to public assessment, it is essential to identify potential developments at an early stage. Within medical science, for example, any one of numerous research lines now being doggedly pursued could create considerable social problems. Within a few years, it may well be possible for parents to choose the sex of their offspring. Will it then be necessary for governments to regulate sex ratios? Is such a capacity one that we wish to have generally available anyway? At the other end of the human life-span, there are now solid expectations that drugs or special dieting techniques, which increase appreciably the life span of rats, will prove applicable to man, and that it will be possible to extend active life.[24] Yet, in a world threatened by population pressure and leisure problems, an effective anti-ageing pill could spawn prolific social, domestic, political and legal dilemmas. Above all, who should have the new elixir? Everyone? Selected individuals? Should the pill be banned altogether? The problem is made more acute by the tantalising possibility that, by postponing not just ageing but also all those degenerative diseases (such as rheumatism) which accompany the ageing process, the elixir could prove to be a panacea incomparably more cost-effective than all the individual treatments now deployed against the diseases of old age. Despite such enormous potential difficulties, we as citizens have not been informed or brought into democratic discussion of the results of research which we finance and which could have far-reaching repercussions in society.

One reason why scientists tend to be diffident about such debate is that they fear the possibility of more control over science, and pressure to curb research in particular areas. Increased regulation, in some areas, is certainly inevitable and necessary. A typical example is the Swedish Concession Board which was set up in 1969 as part of the most comprehensive environmental legislation anywhere in the world. If a chemical company in Sweden wishes to build a plant for producing some novel chemical – or even to change its process for making an existing product – it must first satisfy the board that the new investment will not cause environmental damage. Legislation of this sort to prevent, rather than combat, problems stemming from science and technology will no doubt increase. We shall also see (*pace* Commoner) manufacturers having to bear the total costs of their own pollution and pollution control.

Some research projects would probably be killed – and rightly so – given wider social evaluation. There is no doubt that if the British public (or even MPs) had been fully aware at the right time of the costs and advantages of being able to travel at the speed of an artillery shell, Britain would not have embarked upon the ludicrous business of building Concorde. Nor would a public conscious of the risks and benefits choose to have enzyme detergents or sophisticated fly-papers. Both are excellent examples of the artificially generated 'needs' of J. K. Galbraith's 'affluent society'. Within research institutions, too, we need much greater flexibility, so that research projects which have reached a dead-end are killed at the right time instead of being allowed to drag on into fatuous obscurity, consuming resources of time, money, and manpower. (We should be much more concerned about the *quality* of science, and quick to purge shoddy research.) On a broader canvas, there is little question too that – however the current arguments between the prophets of doom and the optimists are resolved – we *shall* move towards a phase of social development less obsessed with conventional economic growth, and that this will curb our enthusiasm for everlasting innovation and built-in obsolescence, in favour of recycling and ecological stability.

Another fear is that, by ploughing more research into the eminently practical, at the expense of what used to be called 'pure research', we shall impair the seed-bed from which useful science and technology arise, as well as curbing intellectual excitement for young scientists and reducing the flow of qualified manpower. This is a powerful argument, as long as we require qualified scientific manpower. But we must remember that the evidence for a one-way flow of innovation from pure into applied science now looks a great deal more tarnished than before people had actually examined the evidence (p. 93). Also, the exponential phase of growth of basic science is now drawing to a close. Even concentrating on science of immediate practical utility, however, need not mean a diminution of intellectual vigour. The Ph.D. student, for example, can be given a problem that is both intellectually stimulating and socially important – rather than one he can boast of as having no practical significance.

It is entirely possible that one effect of a wider public debate might be to demand more science, not less. As we saw earlier, most of the activist groups outside science which have concerned themselves with scientific issues (such as 'vivisection' and fluoridation) have been essentially negative. What may well happen more frequently in the future is that organisations will arise which work positively for funds to go into particular areas – just as permanent and *ad hoc* groups now lobby for medical services, such as better kidney machine facilities, or money to build a new hospital. Before describing some of the likely topics for positive lobbying, there is one vital consideration in assessing new institutions and changes in the existing organisational fabric of science designed to render science more sensitive to human need and to democratise the decision-making process. This is the urgent importance of maintaining a totally independent body of scientific expertise, as found at present in the universities. Curiously, the need to safeguard academic independence and integrity is an issue on which those critics who attack the increasing involvement of university scientists in contract research for industry and the military, find themselves in close agreement with the traditional guardians of academic scholarship.

The need to maintain a body of independent scientific judgment stems from the critical temper of science itself. As we saw in Chapter 3, science is rarely a matter of black-and-white reasoning or conclusive experimentation. Scientific truths (i.e. what most scientists believe to be true) emerge by the gradual accretion of new information and the open assessment of rival theories. Judgment is just as important here as in business or politics. When a company releases a new drug or insecticide on to the market, for example, there can never be conclusive proof of the product's safety. It is launched on the basis of a mass of data, experimental and conjectural, and the judgments of many different scientists about that data – the staff of the company marketing the new substance, members of government advisory bodies who lay down standards, scientists working for statutory agencies who enforce regulations and screen information about new products, and the experts in both categories throughout the world who have vetted similar products in the past.

But such a closed system is not enough. Nor is it, in practice, the only channel of assessment. The scientific community at large also plays an important role, when the scientists involved expose a new product and its properties to critical discussion by publishing papers in journals and speaking at meetings. Clearly, the more open this informal system of evaluation, the better. At an early stage, before a new product has patent protection, some secrecy is inevitable, and (as in the case of the fly-strip mentioned above) the system could and should be improved. What is unquestionable is the essentiality of independent experts in assessing test data and safety standards without any vested interest in particular products. There are countless other examples in other areas of science, from civil engineering to computer science. Hence the importance of university scientists. Unlike their colleagues in private companies and government bodies, they do not usually have contracts which prevent them from expressing their views on controversial questions. Like their academic counterparts in disciplines such as economics, they are particularly useful to society because of their independent judgment, which is not tied to government policy or company profit. In the past, university scientists have been responsible for drawing attention to the side effects of drugs and pesticides and a range of similar hazards. In so doing, independence is vital.

Just as insidious in eroding this independence as the increase of contract research work for industry and the military, is the way in which government grant-awarding bodies can use their powers to curb criticism when it becomes uncomfortable. As the case of the Swedish Nobel Laureate Professor Hannes Alfvén shows,[25] this practice is commoner than one might suppose. Some years ago, the Swedish government cut its research grants to Alfvén's research institute as a direct result of his opposition to plans for the Markiven nuclear plant. He was warned of likely financial reprisals but persisted in his criticism (since dramatically vindicated: Markiven wouldn't work and today is operating as an oil-powered station) and he now spends much of his time in the United States. 'Society is obviously in serious danger when scientists are subjected to strict and uninformed tutelage,' Alfvén wrote. 'The demand that they have greater freedom to speak out should not be regarded as an attempt to secure privileges for them, but as a recognition that our highly complex society cannot function if a large number of its foremost experts are muzzled.' After those words were published, the Swedish government apparently succeeded in preventing Alfvén from lecturing at the UN Stockholm conference on the human environment in June 1972, on the grounds that his proposed speech – on his doubts about safety standards in nuclear reactors – was too controversial.

Among a long list of social problems which cry out for *more* scientific analysis are such questions as the influence of television violence on the attitudes and behaviour of viewers, especially the young, and the alleged adverse effects of pornography. Both are socially important issues amenable to rigorous scientific investigation. One does not have to be a blind enthusiast for science to see the absurdity of a situation in which one American president (Johnson) sets up an expert-packed National Commission on Obscenity and Pornography and another (Nixon) rejects its conclusion (based on two years of intensive research) that pornography does not lead to crime, on the grounds that *he knows* that pornography can corrupt people. Clearly, a little of Waddington's 'scientific attitude' would help here.

Then there are urgent needs in the area of community science. The British Society for Social Responsibility in Science, for example, during the debate on the Rothschild report in 1972, discussed the need for what it called Community Science Resource Councils, to provide scientific knowledge and expertise for minority and under-represented groups in the community wishing to challenge regional or government policy decisions affecting them. On the world scale, there are such major challenges as ameliorating the effects of earthquakes and setting up crisis squads, using scientific methods and equipment, to deal with global disasters. Problems such as modifying the ill effects of earthquakes (or, though less likely, predicting them) are amenable to scientific research, and the extent to which they succumb is largely a reflection of the number of man hours of research put into them.

Whatever else they require in terms of political initiatives and changed attitudes, the evils for which science and technology are most frequently under attack these days also depend upon science and technology for their solution. To establish and enforce regulations against pollution, we need monitoring equipment to measure the levels of particular chemicals in the air and water. Disarmament and arms control devolve around sophisticated techniques of verification. The prospects of a comprehensive nuclear test ban, for example, depend largely upon exquisitely sensitive techniques for distinguishing underground nuclear explosions from earthquakes. And in the long-term, one hopes that insights from the science of 'peace research' will help us to curb man's propensity to make war at all.

Even the very data which the doomsters use to describe our plight come from advanced techniques of computer modelling. Such methods are still crude (the chief reason why the early work of Meadows and his colleagues has been so heavily criticised), but their potential contribution in helping us to forestall and anticipate trouble, rather than try to deal with disaster when it occurs, is unquestioned. Futurology, or future forecasting, is now an established science with its own journals, including *Futures* in

Britain and *The Futurist* in the USA. This new discipline could become even more important by exposing areas needing scientific research, rather than (as at present) merely concentrating on the passive analysis of future trends.

The potential contribution of science to the developing countries is immense, though again a change of will is necessary, and political initiatives beyond the scope of this book. As we saw earlier (p. 117), there are differing views about how best to harness science to the needs of the Third World.[26] Most recently, now that the industrial West has realised the cost it pays for its scientific and technological development, there are strong doubts as to whether the Third World is really well advised to follow the same path. Even DDT, which prevents some 100 million cases of malaria every year in India, is now under attack, because we have recognised its serious effects on wildlife. Even Norman Borlaug's 'green revolution' is in question, blamed for everything from disturbing the economic pattern of agriculture in developing countries, to boosting the population explosion of both men and grain-consuming rats, and encouraging the proliferation of insect pests. Yet there can be no doubt that science has made enormous contributions to the conquest of poverty and malnutrition, and that more science is required; the failure is a failure to consider its use in relation to ecological stability, and with regard to political and social structure.

Hence to the concept of 'intermediate technology', and a somewhat unexpected consequence of contemporary environmental and scientific debate. Intermediate technology is a halfway house between the primitive and the sophisticated – it is markedly cheaper in capital per man than the advanced technology of industrialised countries, yet gives greater output per man than the craftsman working without any modern aids whatever. A large automated petrochemical works in the West, for example, can be so capital intensive that only one man is required to control 50,000 dollars-worth of equipment. In a developing country, where capital is scarce and labour cheap – the reverse of the case in the developed nations – the most appropriate technology is often intermediate between the two extremes. But politicians in the developing countries have tended to suspect such a halfway house, seeing it as a means of delaying national development rather than encouraging it. Now things have changed. We have realised that pollution and many of our other problems stem in part from our habit of forcing technical ingenuity to its ultimate, costly limit, in the fanatical pursuit of growth and progress, and of fabricating technology on the giant scale. If we can demonstrate, on long-term environmental grounds, the desirability of what has been labelled soft technology (see p. 201), which resembles intermediate technology at many points, then this will go far towards its advocacy in the Third World.

It is, of course, fashionable in some circles to suggest that the type of questions discussed above are all short term and partial: that instead of nuclear test verification, we want an end to war; that instead of putting filters on car exhausts or taking lead out of petrol, we should abolish the automobile; that rather than build more power stations, we question the basis of a society which wastes energy on luxury comforts and entertainments; that developing countries would be advised to remain in their 'more natural' state. Such pontification usually comes from those who have taken for granted the staggering benefits of science, in medicine, agriculture, communications, and other fields and are happy to continue to do so. (As with agnostics living off the moral capital of an apparently discarded Christianity, such judgments resemble the inanities of fans at a pop festival a few years ago, ferried there by technology, clothed by polymer science, fed by food technology, supported by medical science in the event of appendicitis or tetanus, listening to commercial music amplified by electronic wizardry and telling everyone how marvellous it was to create an alternative society.) We should, of course, question continuously the sort of society we have created, and consider radical changes, but a total redesign of society and men's minds is simply not on the agenda for the foreseeable future. Meanwhile there are formidable problems that require science to solve them, and scientific choices to be made. Far more important than intellectual speculation about a New Jerusalem is to ensure that the community at large has proper access to the decisions that are taken.

Recent progress and future trends

Are we moving in the right direction, towards greater democracy in decision-making about science and technology? This question can be resolved into two parts: Are we collecting and digesting the right sort of information on which to make sensible judgments? Secondly, is the information available at the right time to the right people? On the first score, there have been a number of hopeful developments in recent years. One is the emergence of the practice of 'technology assessment', which aims to anticipate, on an independent, unfettered basis, the effects, good and bad, of new science and technology. In 1972 an International Society for Technology Assessment came into being and the *Journal of Technology Assessment* was launched. A recent example of the new discipline was a study at Cornell University of the likely impact of mass-produced microwave devices, which will be coming into use for communications in the next decade or so. Used to facilitate telephone installation in cars and direct satellite-to-earth communication, to detect crime, and for many other purposes, they could proliferate as widely as television in the next few years. Yet there are associated dangers, one of which is that invisible microwave radiation could constitute a new form of pollution, unseen and undetected yet potentially dangerous. Also, some experts believe that microwave devices used as telephone links would be open to eavesdropping; others argue that they will make confidential communications more secure. It was to stimulate the scientific community to consider these and other broad social implications of microwave technology that two members of the Cornell group published an important article on the subject[27] early in 1972.

The idea of technology assessment was first popularised by Congressman Emilio Daddario when he was chairman of the House-Sub-committee on Science, Research and Development of the US Committee on Science and Astronautics. He introduced a Bill in 1967 to create an Office of Technology Assessment for Congress, but it was not until February 1972 that the House of Representatives approved his plan. After a period of virtual eclipse, the Bill was revived by Congressmen frustrated in obtaining technical information, particularly since the Nixon Administration took over and became embroiled in battles with Congress over such issues as ABM deployment and the proposed SST programme. In Britain, there is no official body concerned with technology assessment, though a few university departments have research directed towards costing the hazards of technology. One, for example, at Sussex University, has been concerned with risk assessment in industry. Clearly, there is room for considerable development at this work. There is also great promise in another recent development – the occasional practice of research workers in putting price tags on their discoveries – the cost per patient of a newly feasible life-saving technique, for example, or the financial costs and benefits of pollution abatement. Information of this sort is an essential prerequisite for discussion and debate.

Turning to the institutions and fora in which such information can be evaluated, the United States has a more open and sensitive system of debate than is the case in Britain. With the executive arm of government separate from the legislature, there is greater public scrutiny of government activity. Also, Washington politics have long been less opaque than those of Whitehall, and the USA has the excellent institution of Congressional hearings, at which scientists and politicians concerned with science can be put on the spot and asked awkward questions about their decisions and their research. Britain followed suit in 1966 with the setting up of the all-party Select Committee on Science and Technology. On frugal resources of money and research facilities, it has tackled subjects such as coastal pollution, the computer and nuclear industries, and the implications of the Rothschild report. The Committee calls witnesses to its public hearings and produces reports which, though not usually debated in the House of Commons, have had an important influence on policymaking on technical and scientific matters. The British parliament has also introduced Green Papers (as opposed to White Papers announcing firm policy) which describe unfinished government thinking as a basis for open discussion. The Rothschild report, for example, appeared in a Green Paper.

Scientific societies, too, although they have a long way to go before they become involved in the cut and thrust of political action, have begun to hold meetings at which social and political implications of science are debated. Britain's Institute of Biology, for example, which used to devote its annual symposia to such topics as 'Freezing and Drying', 'Biological Aspects of the Transmission of Disease', and 'The Biology of Deserts', has in recent years turned to urgent social questions. Subjects tackled recently include 'The Optimum Population of Britain' (1970), and 'The Future of Man' (1969). In the United States, the New York Academy of Sciences has held conferences on such subjects as 'Environment and Society in Transition' (1970) and 'The Social Responsibility of Scientists' (1971). At worst, such conferences can be merely wordy gestures towards topics of burning concern – just as some industrial companies have taken to organising meetings, essay competitions, and all manner of other gimmicks as their contribution towards environment, conservation and ecology. At best, they are worthwhile and represent genuine concern. A recent example of grandiloquent impotence was the Weizmann Institute's symposium on 'The Impact of Science on Society', in Brussels in June 1971. At the same time, the Ciba Foundation in London was holding its excellent symposium entitled 'Civilisation and Science: in Conflict or Collaboration?', which I have referred to several times in this book.

Consultation and democratic evaluation may, it seems, be improving. According to those arch-critics, the Roses,[28] even the presentation of the arguments and the decision-making procedure for the British government's 1968 decision on the CERN accelerator were a substantial improvement on a directly comparable earlier case, the UK debate about a proposed international research centre in Edinburgh for the World Health Organisation. This proposal was finally defeated, after four years of relatively secretive debate, at the WHO 18th World Health Assembly in Geneva, in 1965. But, as we have seen from the missing links in the 1968 CERN saga, there is still a considerable way to go. So it is with the availability of information. In 1970, the US Environmental Protection Agency decided to make available to all *bona fide* inquirers the submissions made by manufacturers about new pesticides – an important step in exposing such data to independent assessment by the (international) scientific community. In Britain, such information is still secret.

One ominous development recently is the closing down or curbing of a number of exercises set up specifically to assess in depth the problems of science in modern society. Harvard University's 'Programme on Technology and Society', for example, was ended prematurely in June 1972, after eight years during which it had published important analyses of socio-technological issues and made increasing inroads in understanding the social impact of science. At about the same time, Columbia University closed down its Institute for the Study of Science in Human Affairs. In each case, internal institutional difficulties were responsible, rather than any realisation that the problems were too complex or insufficiently important to pursue. On the contrary, work at both Harvard and Columbia emphasised just how far-reaching scientific and technological change can be in society, and exposed the need for greater democratic control.

It is, of course, easy to list examples (as in Chapter 8) showing how the citizen has played a more active role in recent years in making his views felt and in influencing scientific and technological matters. The public and its elected representatives killed the American SST programme in 1971. Public concern, mobilised and channelled by the Press, pressured a group of distinguished British doctors to visit Dr Issels' Ringberg-Klinik early in 1971, and helped to curb surgeons' enthusiasms for heart transplantation in both the United States and Europe. A vigorous local campaign (backed by big money) made the British government change its mind after an 'irrevocable' decision by a second expert committee on London's new airport. It was partly public and political pressure too which persuaded the US National Science Foundation, in 1972, to begin to spend some of its funds, formerly totally allocated to 'pure science', on mission-orientated research. One can even argue, as Anthony Wedgwood Benn does,[29] that in a modern, highly interdependent science-based community the citizen has more power, not less, to disrupt its harmonious workings. There will doubtless be many more incidents like the one at the Swansea carbon black factory, many more (and increasingly sophisticated) attacks on companies such as RTZ. But what governments in particular have almost totally failed to do as yet is to provide an adequate official framework for democratic discussion of science. One searches in vain for hopeful signs. I have found just one –

early in 1971, at West Germany's Nuclear Research Centre at Karlsruhe, with the Minister of Science present, a *public* debate took place on the development of fast breeder reactors in the Federal Republic. Expert argued with expert, and the aim was to make 'the right of the public to have such large-scale projects as transparent as possible'. The event was probably unique not only in West Germany but in any Western country.

In fact, the changes that have occurred towards greater democratisation have all been forced upon governments, just as the Swansea housewives persuaded the carbon black factory to scrap its expansion plans. In their turn, politicians, disappointed by the failure of science to live up to its economic promise, have put pressure on the scientific community. By and large, the response of scientists to unaccustomed criticism and questioning has been ill-tempered (the 'you can't possibly understand' attitude again) or apathetic. In spite of the movements described in the last chapter, rank-and-file working scientists remain unaffected by calls to social responsibility and uninterested in the wider context of their work. In his memoirs[30] Vannevar Bush, a former high mandarin of US science and defence policy, writes as though he never had any doubts that science is entitled to special public support, and that technical and scientific decisions, whatever their effects on our lives, must be left to the wise men who know, without any genuine democratic control. It is not a consciously wicked attitude, but simply the result of a rigid and totally confident view of the scientist's role in society. Couched in less flamboyant and robust terms than those used by Bush, subconscious élitism is found in the majority of working scientists. This attitude will have to change, as Jerry Ravetz argues, so that scientists no longer separate the professional from the social. They must be willing to expose their own work to wider democratic influence, and become more active themselves politically and socially. Against this background, it makes considerable sense for Chairman Mao to have sent his scientists to work in the fields and factories, where they can learn at first-hand about everyday problems. Similarly, it was right for not only the Chinese delegation, but also the Swedish Premier Olaf Palme and the Indian Prime Minister Mrs Indira Gandhi, to attack United States ecocide in Indochina, at the UN environment conference in 1972. To pretend that the ecological ravages of the Vietnam war are somehow different to other environmental damage would be hypocrisy. The opposite approach, based on rigid compartmentalisation of science, is exemplified by the thousands upon thousands of American scientists who knew that the fall-out shelters built in the late 1950s would be ineffective, yet did nothing about it. When the soundly-based Scientists' Committee on Radiation was set up to inform the public about the shelters and radiation dangers, several important scientific societies refused to join or support it. Scientists believed they had done their job by collecting, analysing, and writing up the relevant information. It was up to others to decide what action to take.

The recognition that science is not neatly separable from its social context, nor pure science divisible from that which is applied, has led to the suggestion of an oath for scientists, similar to the Hippocratic oath of the medical profession. The following formulation was designed for applied scientists, engineers, and technologists by Professor Meredith Thring of Queen Mary College, London:

'I vow to strive to apply my professional skills only to projects which, after conscientious examination, I believe to contribute to the goal of co-existence of all human beings in peace, human dignity and self-fulfilment.

'I believe that this goal requires the provision of an adequate supply of the necessities of life (good food, air, water, clothing and housing, access to natural and man-made beauty), education, and opportunities to enable each person to work out for himself his life objectives and to develop creativeness and skill in the use of hands as well as head.

'I vow to struggle through my work to minimise danger; noise; strain or invasion of privacy of the individual; pollution of earth, air or water; destruction of natural beauty, mineral resources and wildlife.' (from M. W. Thring, *New Scientist*, 7 January, 1971)

Such an oath would certainly help to focus scientists' attention, at the outset of their careers, on their social responsibilities. It would also emphasise that it is not only medicine and medical science which raise ethical dilemmas and social problems. In the last few years the moral issues behind such questions as the use of lobotomy and other surgical operations to treat mental illness, the determination of the 'moment of death' (particularly its relevance to resuscitation procedures and the selection of donors for organ transplants), and abortion in cases of confirmed congenital disease, have been discussed in the press. But what of the chemists who helped to devise the improved, stickier napalm? What of the optics man designing infra-red equipment to be used in night warfare in Vietnam? What of the physicists who exploded the 'Rainbow bomb' and those who exploited the bandwaggon of Mohole? There is even a current textbook, entitled *Ethics for Scientific Researchers*,[31] which is concerned almost entirely with medical science. A young physics undergraduate, picking up the book, could be forgiven for assuming, with relief, that his own discipline is comparatively free of ethical dilemmas. He would be wrong, and anything that can be done to direct the attention of all young scientists towards their ethical and social responsibilities is to the good.

It simply is not an acceptable answer to argue that because science sometimes has unforeseen effects, the scientist cannot be held responsible for his work. In a few cases, this is true. The German chemist who first evolved polystyrene in the last century had no means of knowing that it would later be used to make 'superior' napalm. But in many cases such developments *can* be foreseen, and certainly scientists have an obligation to ponder them. The chemist who, while developing insecticides, chanced upon the precursor of the horrifyingly deadly nerve gases, could have predicted possible applications of his discovery. And when an unexpected, unforeseeable calamity does happen, the scientist is *more*, not less or equally, responsible compared with others in doing all he can to fight against it. Arthur Galston is the exemplar here.

Nowhere is the importance of searching one's own conscience more crucial than in military work. Something like 15 billion dollars have been spent each year throughout the world on military R and D over the past decade (compared with about four billion dollars annually on medical research). And the only tangible achievements of the SALT talks and other efforts at arms control have been agreements to limit or destroy already obsolete weaponry. It is tempting, therefore, to be totally pessimistic about curbs on the arms race. In a book of this sort, all considerations of the social relations of science can easily be swamped by the massive deployment of science for the military. Yet the fact remains that modern warfare depends on science, and if the 20 per cent of the world's scientists who now work in defence were to turn their talents to peaceable ends, this would be a most potent way of curbing military escalation and aggression. For the young scientist, the excellent facilities which defence research always offers can be a powerful, seductive attraction. For scientists of all ages, contract work for the military provides extra cash, equipment and prestige. While one cannot quibble with the scientist who works in defence for conscientious reasons, those who are merely tempted by money and facilities should consider all the implications of what they are doing and resist seduction. There are signs that young scientists in particular are beginning to do this. There is hope too in the development of something like an open debate on ABM deployment in the USA in recent years. Such public discussion, and the resulting small margin of success for the anti-ABM lobby, could be a harbinger of similar movements in the future.

Returning to civil science, what we do not want, except in the most extreme circumstances, is that other panacea often advocated by protagonists of social responsibility: the enforced moratorium on research. Such a solution was proposed following Arthur Jensen's work published in 1969, on the mental abilities of Negro and white Americans. It was felt that any research of this sort was so dangerous politically that it should be stopped. This is a naive assessment. Comparative studies of mental skills *are* important and necessary if educationalists are to adjust teaching methods to take advantage of any special characteristics in different ethnic groups. Clearly, any such decisions should be agreed between the different groups involved: they should not, for example, be taken by a black or white majority in a community on the basis of research conducted by that majority. Moreover, just as ability in such skills as mathematics or languages varies from person to person in any school class, this does not imply that any one group is superior to another. Who is foolish enough to claim that the typical Western IQ test is an objective universal arbiter? Who will rate IQ higher than other human qualities? What we *can* say is that in one context – when IQ tests are used to 'rescue' bright children in an educational system where their talents would otherwise atrophy – IQ tests are a progressive instrument; in another context, they are condemned as a weapon of élitist meritocracy. Yet the alternative to soundly-based research on intelligence by responsible scientists, exposed to open evaluation, is either continued ignorance (and thus continued prejudice and suspicion – the very worst brand of political dynamite) or research carried out by less capable or less scrupulous individuals.

In any case, while it is practicable to proscribe an area of costly and highly sophisticated research – in nuclear physics, for example – it is scarcely realistic to talk of preventing psychologists from conducting intelligence tests. Though there will be increasing public pressure behind the idea unless scientists mend their ways, a moratorium is rarely a satisfactory or even feasible answer, and is usually a way of avoiding social or political problems which are bound to recur, perhaps in a more acute form. In May 1972, for example, the *Journal of the American Medical Association*[32] called for a moratorium on attempts to implant into a woman's womb an ovum fertilised outside the body, largely on the grounds that it is 'not a proper goal of medicine' to enable women to have children by a means which, being new, carries some risk and uncertainty. Surely the real answer is to get down to solving the moral and technical problems raised by this newly feasible technique, rather than seek to ban entirely a method of helping otherwise infertile women to have children?

Perhaps the most effective means of stirring the conscience of the working scientist is by educating the would-be scientist into an awareness of the wider implications of his subject. As we have seen, the training of the scientist is still usually a narrow one, which ignores the place of science in society and even fails to consider critically the scope and assumptions of science itself. We could do much here to call the bluff of science (as Polanyi, Roszak, and others have done in their very different ways) and much in education generally to place the non-scientist's hopes and fears of science in realistic perspective. Again there is also a positive side: would-be politicians and administrators desperately need *greater* scientific understanding and literacy so that they can assess scientific needs, claims, projects, priorities, threats and promises, with more understanding and less ignorance and inhibition.

Looking to the future, I hope we shall live to see not the abolition of science (as some would like), nor the disestablishment of all science (Bronowski-style), and certainly not a resurgence of blind, whizz-bang adulation, but a gradual resolution of science into its two poles: the pursuit of knowledge for its own sake, and the conquest of tangible practical problems. We may see scientific mercenaries, working in the mission-oriented laboratories of industry and elsewhere – including new institutions devoted to social problems – and on the other hand scientists who, like poets, pursue science because they must. Society will both support science as an essential cultural and intellectual activity, and exploit science for its hard practical returns. At present these two motives are hopelessly confused – and that is the source of much of our contemporary disquiet.

Notes

CHAPTER ONE

1 Quoted in *Science and the Mass Media* by Hillier Krieghbaum, New York University Press, 1967, p. 5
2 Public Affairs Press, Washington DC
3 Cambridge University Press, 1957, p. 145
4 Macmillan, 1970
5 by Gordon Rattray Taylor, Thames and Hudson, 1970
6 edited by G. R. Urban, The Bodley Head, 1971
7 edited by Jack D. Douglas, Prentice-Hall Inc, 1971
8 by Hans J. Morgenthau, New American Library, 1972
9 vol. 167, 1970, p. 141
10 *Science*, vol. 174, 1971, p. 2
11 *Science*, vol. 176, 1972, p. 990
12 *New Scientist*, 30 September, 1965, p. 849
13 *New Scientist*, 10 February, 1972, p. 340
14 *Harvest of Death*, by J. B. Neilands and others, Collier-Macmillan, 1972
15 *Sunday Times*, 9 July, 1972
16 *Scientific American*, May, 1972, p. 21
17 *Radio Times*, 24 October, 1971
18 12 April, 1972, p. 486
19 *Science*, 2 February, 1973, p. 459

CHAPTER TWO

1 Pelican, 1969, p. 15
2 Quoted from *A Short History of Science*, by J. G. Crowther, Methuen, 1969, p. 179
3 Preface to *Where is Science Going?* by Max Planck, Allen and Unwin, 1933
4 Hutchinson, 1964
5 Walter Scott, 1892, p. 7

6 *The Listener*, 12 October, 1967
7 Translated into English as *The Logic of Scientific Discovery*, Hutchinson, 1959
8 *Scientists at Work*, edited by Tore Dalenius, Georg Karlsson, and Sten Malmquist, Almqvist and Wiksell, 1970, p. 243
9 Reprinted in *The Listener*, 25 August, 1960
10 Hamish Hamilton, 1966, p. ix
11 *New Scientist*, 2 November, 1961, p. 306
12 Allen and Unwin, 1970
13 *The Art of Scientific Investigation*, by W. I. B. Beveridge, Heinemann, 1961, p. 167
14 Pelican Books, 1941
15 Detailed in *Induction and Intuition in Scientific Thought*, by P. B. Medawar, Methuen, 1969, p. 3
16 *Design of Experiments*, by Sir Ronald Fisher, Oliver and Boyd, 1935
17 *As I Remember Him*, by Hans Zinsser, Macmillan, 1940, p. 140
18 *Obras Completas*, vol. 6, 1958, p. 143

CHAPTER THREE

1 In *The Sociology of Science*, edited by B. Barber and W. Hirsch, Collier-Macmillan, 1962, p. 389
2 Cambridge University Press, 1968, p. 9
3 Honor Fell, *The Advancement of Science*, vol. 27, 1970, p. 129
4 2nd edition, University of Chicago Press, 1970
5 N. R. Hanson, *The Concept of the Positron*, Cambridge University Press, 1963
6 Williams and Norgate, 1950, p. 33
7 *The Social Process of Innovation*, Macmillan, 1972
8 Weidenfeld and Nicolson, 1968; Penguin Books, 1970
9 *The Hope of Progress*, Methuen, 1972, p. 105
10 *Journal of General Microbiology*, vol. 63, 1970, p. 1
11 See *The Scientific Community*, Warren O. Hagstrom, Basic Books, 1965
12 *Little Science, Big Science*, Columbia University Press, 1963, Paperback edition, 1969, p. 50
13 J. Gaston, 'Big Science in Britain: A Sociological Study of the High Energy Physics Community', Ph.D. thesis Yale University, 1969
14 *Experiment*, edited by David Edge, BBC Publications, 1964
15 *The Listener*, 2 September, 1965, p. 345
16 *Journal of Biological Chemistry*, vol. 193, 1951, p. 265

17 See, for example, Ailon Shiloh, *New Scientist*, 26 April, 1962, p. 169
18 *The Art of the Soluble*, P. B. Medawar, Pelican, 1969, p. 128
19 John Martyn, *New Scientist*, 6 February, 1964, p. 338
20 Anthony Tucker, *The Guardian*, 4 May, 1965
21 *Impact of Science and Society*, vol. 21, 1971, p. 151
22 *Lancet*, 1968, vol. 2, p. 1049
23 *New Scientist*, 26 February, 1970, p. 402
24 vol. 238, p. 198
25 *The Piltdown Fantasy*, Francis Vere, Cassell, 1955
26 *On Understanding Science*, Mentor Books, 1951, p. 23
27 *Minerva*, vol. 8, 1970, p. 324

CHAPTER FOUR

1 *Minerva*, vol. 1, 1962, p. 54
2 Ashby, E., *Community of Universities*, Cambridge University Press, 1963, p. 16
3 *The Teaching of Science*, Allen and Unwin, 1969, p. 140
4 *Science, Industry and Society*, by Stephen Cotgrove and Steven Box, Allen and Unwin, 1970, p. 63
5 *The Struggles of Albert Woods*, W. Cooper, Jonathan Cape, 1952
6 *Graduate Education: Parameters for Public Policy*, National Service Board, Washington, 1969, p. 23
7 *Louis Agassiz: A Life in Science*, University of Chicago Press, 1960
8 *Graduate Education in the United States*, McGraw-Hill, 1970
9 'Science and the Common Reader', *Commentary*, June 1966
10 *National Research Council News Report*, vol. 21, no. 4, 1971, p. 2
11 *Nature*, vol. 228, 1970, p. 813
12 *Nobel, The Man and his Prizes*, H. Schück *et al*, Elsevier, 1962
13 *Scientific American*, November 1967; *American Sociological Review*, vol. 32, 1967, p. 391
14 Hodder and Stoughton, 1971, p. 331
15 *The Politics of American Science*, Penguin Books, 1969, p. 38
16 In *The Sociology of Science*, edited by B. Barber and W. Hirsch, Collier-Macmillan, 1962, p. 201
17 *Contrary Imaginations*, Shocken, 1966
18 *The Advancement of Science*, vol. 27, 1970, p. 6

CHAPTER FIVE

1 J. D. Bernal's *The Social Function of Science*, first published in 1939, is still the best guide. The edition published by MIT Press in 1967 contains a valuable essay written by Bernal 'After Twenty Five Years'
2 'A time to think', *The Advancement of Science*, vol. 27, 1970, p. 1
3 *The Environment Crisis*, edited by H. E. Helfrich, Yale University Press, 1970, p. 164
4 Royal Society of Arts Trueman Wood lecture, 1971
5 See, for example, R. R. Porter, *Presidential Addresses*, British Association, 1971
6 See *The Pill on Trial*, Paul Vaughan, Penguin, 1972
7 Quoted in *The Body*, Anthony Smith, Allen and Unwin, 1968, p. 110
8 See *The Green Revolution*, by Stanley Johnson, Hamish Hamilton, 1972
9 *The Universal Eye*, Timothy Green, The Bodley Head, 1972
10 *Science Journal*, October 1969, p. 44
11 *The Science of War and Peace*, Robin Clarke, Jonathan Cape, 1971, p. 181
12 Steven Rose and Hilary Rose, *New Scientist*, 21 January, 1971, p. 134
13 In *The Pentagon Papers* and *Science*, vol. 176, 1972, p. 1216
14 *Chemical and Biological Warfare*, edited by Steven Rose, George Harrap, 1968, p. 62
15 *Efficiency in Death*, Council on Economic Priorities, Harper & Row, 1970
16 *New Scientist*, 30 March, 1972, p. 685
17 *Unless Peace Comes*, edited by Nigel Calder (Allen Lane The Penguin Press, 1968) provides more glimpses into the horrifying weaponry of the 1980s and beyond
18 See, for example, Steven Rose and Robert Smith, *New Scientist*, 4 September, 1969, p. 468
19 Cmnd 4775
20 *Science*, vol. 176, 1972, p. 1400
21 J. Langrish, *Science Journal*, December 1969, p. 81
22 Joseph P. Martino, *The Futurist*, April 1972, p. 70
23 M. Gibbons, S. Metcalf, and D. Watkins, *Science Journal*, December, 1970, p. 71

CHAPTER SIX

1 Cmnd 4814
2 M. F. Perutz, *Nature*, vol. 235, 1972, p. 191
3 *The Times*, 31 December, 1971, p. 9
4 27 February, 1971, p. 15
5 The DSIR disappeared under the Labour government's reorganisation of civil science in Britain in 1964
6 *The Sunday Times*, 20 November, 1966, p. 12
7 *The Science of War and Peace*, Robin Clarke, Jonathan Cape, 1971
8 See *The Last Resource*, Tony Loftas, Hamish Hamilton, 1961
9 *Scientists and War*, Sir Solly Zuckerman, Hamish Hamilton, 1966, p. 162
10 *The Social Function of Science*, MIT Press, 1967, p. 170
11 *New Scientist*, 28 December, 1967, p. 746
12 *Race to Oblivion*, Simon and Schuster, 1971
13 *New Scientist*, 24 September, 1970, p. 618
14 *Science and Technology in British Politics*, Norman J. Vig, Pergamon, 1968
15 Harry G. Johnson, *Minerva*, vol. 10, 1972, p. 10
16 vol. 21, 1971, p. 105
17 *New Scientist*, 28 January, 1971, p. 183
18 *New Scientist*, 8 October, 1970, p. 75
19 Amply described in *The Battle for Concorde*, John Costello and Terry Hughes, Compton Press, 1971
20 Sir Ernst Chain, 'Thirty years of penicillin therapy', *Proc. R. Soc. Lond. B*, vol. 179, 1971, p. 293
21 *The Sources of Invention*, John Jewkes, David Sawers, and Richard Stillerman, Macmillan, 2nd edition, 1969, p. 353
22 *Science*, vol. 173, 1971, p. 259; *New Scientist*, 16 October, 1969, p. 122
23 *New Scientist*, 2 February, 1967
24 Edwin Mansfield, *Science*, vol. 175, 1972, p. 485
25 J. Langrish, M. Gibbons, W. G. Evans, and F. R. Jevons, Macmillan, 1972
26 *New Scientist*, 27 January, 1972, p. 223
27 *Success and Failure in Industrial Innovation*, Centre for the Study of Industrial Innovation, London, 1972
28 'The discovery of polythene', R.I.C. Lecture Series 1964, No. 1, R. O. Gibson
29 Callum McCarthy, *New Scientist*, 27 February, 1969, p. 460
30 *Impact*, vol. 20, 1970, p. 183

CHAPTER SEVEN

1 July 21, 1969
2 *Selectivity and Concentration in Research*, Science Research Council, London, 1970
3 *Chemistry in Britain*, April 1970, p. 143
4 October 27, 1964, p. 8. For a fuller account of Mohole see Dan Greenberg's *The Politics of American Science*, Pelican Books edition, 1969, p. 219
5 An oceanographer from Scripps Institution of Oceanography, University of California
6 *Drilling for Scientific Purposes* (Report of a symposium held by the Geological Survey of Canada), 1966, p. 66
7 *New Scientist*, 2 September, 1971, p. 508
8 *New Scientist*, 2 September, 1971, p. 511
9 *Science Journal*, March 1969, p. 4
10 Flowers received a knighthood in 1969
11 'The proposed 300 GeV accelerator', Cmnd 3505
12 *New Scientist*, 18 June, 1970, p. 574
13 May 22, 1970
14 See *Science and the Mass Media*, Hillier Krieghbaum, New York University Press, 1967, p. 40
15 *Science*, vol. 167, 1970, p. 141; Greenberg now publishes the Washington-based newsletter *Science and Government Report*.
16 *Minerva*, vol. 9, 1971, p. 474
17 *Nature*, vol. 206, p. 579
18 *Science Journal*, April 1968, p. 71
19 See, for example, *New Scientist*, 5 October, 1967, p. 33
20 *Antibiotics in Animal Husbandry*, Office of Health Economics, 1969
21 *New Scientist*, 28 March, 1968, p. 678
22 Report of the Joint Committee on the use of antibiotics in animal husbandry and veterinary medicine, Cmnd 4190

CHAPTER EIGHT

1 *The Private Papers of Henry Ryecroft*, Archibald Constable, 1903
2 A detailed account of the campaign appears in *The Price of Amenity*, by Roy Gregory, Macmillan, 1971, p. 133
3 *New Scientist*, 25 February, 1971, p. 406
4 *New Scientist*, 7 August, 1969, p. 292

5 6 March, 1971, p. 516
6 24 November, 1970
7 See *The Dark Face of Science* (John Vyvyan, Michael Joseph, 1971) for an account of the history of the anti-vivisection movement in this century
8 Alwyne Wheeler, *Science Journal*, November 1970, p. 28
9 Michael Allaby's *The Eco-activists* (Charles Knight, 1972) describes some contemporary activist groups concerned with the environment
10 Hamish Hamilton, 1963
11 *The Question Mark*, Collins, 1969
12 See *The Nuclear-Power Rebellion*, Richard S. Lewis, Viking Press, 1972 and *The Atomic Establishment*, H. Peter Metzger, Simon & Schuster, 1972
13 *Nuclear Engineering International*, June 1972, p. 461
14 *Civilisation and Science*, ex Ciba Foundation Symposium Associated Scientific Publishers, 1972, p. 113
15 'The changing pattern of research in economic entomology' by Harry Rothman, *Scientific World*, 1969, no. 1, p. 11
16 *New Scientist*, 29 April, 1971, p. 272
17 See *Drugs, Doctors, and Disease*, by Brian Inglis, Andre Deutsch, 1965 and *Thalidomide and the Power of the Drug Companies*, by Sjöström and Nilsson, Penguin, 1972
18 *Chemical and Engineering News*, 26 July, 1971, p. 24
19 Grossman, 1970
20 *New Scientist*, 30 July, 1970, p. 232
21 Ballantine/Friends of the Earth, 1971
22 8 December, 1967, p. 34
23 25 February, 1971, p. 421
24 *The Observer*, 26 May, 1968
25 *The Observer*, 28 July, 1968
26 *The University and Military Research*, by Dorothy Nelkin, Cornell University Press, 1972
27 *New Society*, 19 February, 1970, p. 301
28 *Growth, the Price we Pay*, Staples Press, 1970
29 *New Scientist*, 13 August, 1970, p. 343
30 Bodley Head, 1970

CHAPTER NINE

1 *In Civilisation and Science*, Ciba Foundation Symposium, Associated Scientific Publishers, 1972, p. 25
2 *Science*, vol. 125, 1957, p. 480

3 Faber and Faber, 1970, p. 5. See also Roszak's *Where the Wasteland Ends*, Faber and Faber, 1973
4 See *Tongues of Conscience*, by R. W. Reid, Constable, 1969, p. 186
5 *The Technological Society*, Jonathan Cape, 1965, p. 138
6 *The Psychology of Science*, Harper and Row, 1966, p. 56
7 Routledge and Kegan Paul, 1958, p. 286
8 I. T. Ramsay, *Religion and Science*, SPCK, 1964
9 Two recent explorations of this sort are *Towards Deep Subjectivity*, by Roger Poole (Allen Lane, 1972) and *The Cult of the Fact*, by Liam Hudson (Cape, 1972)
10 MIT Press edition, 1967, p. 3
11 *Nature*, vol. 230, 1972, p. 283
12 The Bodley Head, 1968 and 1971
13 See Hector Hawton's contribution to *Journeys in Belief*, edited by Bernard Dixon, Allen and Unwin, 1968, p. 105
14 *Science*, vol. 165, 1969, p. 151
15 *The Times*, 17 May, 1968
16 *The Myth of the Machine*, Harcourt, Brace and World, 1967, p. 62
17 *The Times Literary Supplement*, 23 April, 1970
18 Collins, 1972
19 *Bookworld*, 24 October, 1971, p. 4
20 Edited by Arthur Koestler and J. R. Smythies, Hutchinson, 1969
21 *New Scientist*, 25 September, 1969, p. 638

CHAPTER TEN

1 *Science for the People*, March 1972, p. 4
2 Steven Rose and Hilary Rose, *Impact*, vol. 21, 1971, p. 149
3 History of the Pugwash Conferences, by J. Rotblat, Pugwash Continuing Committee, 1962
4 *New Scientist*, 9 April, 1970, p. 63
5 See 'Defending the Environment – A Case Study', by Dennis Puleston, Brookhaven Lectures Series, No. 104, 1971
6 *Nature*, vol. 237, 1972, p. 8
7 *New Scientist*, 25 February, 1971, p. 436
8 The proceedings, edited by Watson Fuller, are published under this title by Routledge and Kegan Paul, 1971
9 *Harvard Educational Review*, vol. 39, no. 1, 1969
10 Race, Culture, and Intelligence, edited by Ken Richardson and David Spears (Penguin, 1972) presents the anti-Jensen position
11 *New Scientist*, 20 April, 1972, p. 136
12 *New Scientist*, 24 February, 1972, p. 424

13 Sierra Club/Ballantine
14 See *The Optimum Population for Britain*, edited by L. R. Taylor, Academic Press, 1970
15 Jonathan Cape, 1972
16 In *Science and Survival*, Gollancz, 1966
17 *Scientific Knowledge and its Social Problems*, Clarendon Press, Oxford, 1971, p. 424
18 *Low-level Radiation*, by Ernest J. Sternglass, Ballantine, 1972
19 *Poisoned Power*, by John W. Gofman and Arthur R. Tamplin, Rodale Press, 1971
20 Columbia University Press, 1970
21 *The Sciences*, January–February, 1972, p. 9
22 *Nature*, vol. 224, 1969, p. 768
23 vol. 224, 1969, p. 1337.
24 *Evening Standard*, 11 February, 1970
25 *Futures*, vol. 4, 1972, p. 168
26 *Encounter*, July 1971, p. 9

CHAPTER ELEVEN

1 Secker and Warburg, 1964
2 *The Common Sense of Science*, Heinemann, 1931
3 *Science in Universities*, Public Lecture at Nottingham University, 6 March, 1970
4 Collins, 1969. See also *Doom or Deliverance?*, Manchester University Press, 1972
5 *Murderous Providence: A Study of Pollution in Industrial Society*, Rupert Hart-Davis, 1972. Contrast this view with that of Sir Eric Ashby, chairman of the UK Royal Commission on Environmental Pollution, that pollution is 'a temporary malfunctioning of the present economic system' (Speech to *The Times* Environmental Symposium, London, 27 June, 1972)
6 P. B. Medawar, *New Statesman*, 19 June, 1964
7 *The Hope of Progress*, Methuen, 1972, p. 110
8 *The Doomsday Syndrome*, Macmillan, 1972
9 *The Politics of the Environment*, Fabian Tract 412, 1972
10 *The Limits to Growth*, D. H. Meadows, D. L. Meadows, J. Randers, and W. W. Behrans, Universe Books, New York, 1972
11 G. L. Wick, *New Scientist*, 6 August, 1970, p. 276
12 R. Gillette, *Science*, vol. 176, 1972, p. 492
13 See, for example, T. Friedman and R. Roblin, *Science*, vol. 175, 1972, p. 949

14 *Genes, Dreams, and Realities*, Medical and Technical Publishing Co Ltd, 1971
15 See *The Shape of Minds to Come*, by John Taylor, Michael Joseph, 1971
16 *Science, Growth and Society*, OECD, 1971
17 See Jon Tinker, *New Scientist*, 2 March, 1972, p. 489
18 Nicholas Wade, *Science*, vol. 175, 1972, p. 498
19 Jon Tinker, *New Scientist*, 9 March, 1972, p. 530
20 *Civilization and Science*, Associated Scientific Publishers, 1972, p. 114
21 *WHO Technical Report Series*, No. 422, 1970
22 See *Nuclear Politics*, by Andrew J. Pierre, Oxford University Press, 1972
23 Allen Lane, the Penguin Press, 1969
24 See Alex Comfort, *New Scientist*, 11 December, 1969, p. 549
25 *Impact*, vol. 22, 1972, p. 85
26 See, for example, E. F. Schumacher, *Resurgence, Journal of the Fourth World*, vol. 3, 1970, p. 1; and Graham Jones's *Role of Science in Developing Countries*, Oxford University Press, 1971
27 Raymond Bowers and Jeffrey Frey, *Scientific American*, February 1972, p. 13
28 *Science and Society*, p. 233
29 *The New Politics, A Socialist Reconnaissance*, Fabian Tract 402, 1970
30 *Pieces of the Action*, New York: Morrow, 1970
31 by Charles E. Reagan, Charles C. Thomas, 1971
32 vol. 220, p. 721

Index